All About Mars Journeys and Settlement

Copyright Page

This book is copyrighted for 2022

All about Mars Journeys and Settlement

The Living in Space Series Book Two

By Martin K. Ettington

All Rights Reserved USA 2022

ISBN: 9798664404722

Printed in the United States of America

I0480674

All About Mars Journeys and Settlement

All About Mars Journeys and Settlement

This is a 2022 revised version of the book first published in 2020. A lot of new information has become available since then.

A manned mission to Mars has been the dream of humanity at least since the nineteenth century when we first saw details of the surface and thought there might be canals filled with water there.

Here I've looked at the history of unmanned exploration of Mars over the last fifty plus years, proposed missions to Mars, Mars Settlements, and other major issues regarding traveling to and living on Mars. Some proposals have lots of details of proposed scenarios if you want to read all of the engineering and scientific analysis work.

I grew up in the 1960s when every kid in America was fascinated with the Space program and the Astronauts. I also watched not only the Apollo 11 moon landing, but all of the successive trips to the Moon and exploration of the surface. This may be a lot of the reason I became an Engineer, worked as NASA in Houston for several years, and applied to the Astronaut Program myself.

There are some probes which reported life on Mars and then other scientists questioned the results. We are still sending unmanned probes today to try to answer those questions.

All About Mars Journeys and Settlement

Other books by Martin K. Ettington

Spiritual and Metaphysics Books:

Prophecy: A History and How to Guide

God Like Powers and Abilities

Enlightenment for Newbies

Removing Illusions to Find True Happiness

Using the Scientific Method to Study the Paranormal

A Compendium of Metaphysics and How to Guides (Six books together in one volume)

Love from the Heart

The Enlightenment Experience

Learn Your Soul's Purpose Pursuing Enlightenment

A Modern Man's Search for Truth

Use Intuition and Prophecy to Improve Your Life

The Handbook of Spiritual and Energy Healing

Pure Spirituality and God

Memories Before Birth and Reincarnation

Paranormal Abilities and the Yoga Sutras of Patanjali

Mystical and Magical Societies and Practitioners

Important Prophecies of the Future

Longevity & Immortality:

Physical Immortality: A History and How to Guide

The Commentaries of Living Immortals

Records of Extremely Long Lived Persons

Enlightenment and Immortality

Longevity Improvements from Science

The 10 Principles of Personal Longevity

Telomeres & Longevity

The Diets and Lifestyles of the World's Oldest Peoples

The Longevity Six Books Bundle

Long Lived Plants and Animals

A Guide to Longevity Foods, Diets, and Supplements

Science Fiction:

Out of This Universe

The Immortals of the Interstellar Colony

The Mystic Soldier

The Immortality Sci Fi Bundle

Visiting Many Universes

The History of Science Fiction and Fantasy

The God Like Powers Series:

Human Invisibility

Invulnerability and Shielding

Teleportation

Psychokinesis

Our Energy Body, Auras, and Thoughtforms

All About Mars Journeys and Settlement

The God Like Powers Series—
> Volume 1 CompilationThe Yoga Discovery Series:

Yoga-An Ancient Art Form

Hatha Yoga-Helping you Live Better

Raja Yoga-Through the Ages

The Yoga Discovery Package

Business & Coaching Books:

Creating, Paublishing, & Marketing Practitioner Ebooks

Building a Successful Longevity Coaching Business

Why Become a Coach?

The Professional Coaching Success Trilogy

2020-Make Money Writing and Selling Books

The 2020 Handbook of High Paying Work Without a College Degree

The important of Creativity and How to Improve Yours

Quantum Mechanics, Technology, Consciousness, and the Multiverse

Self-Improvement

Stress Relief and Methods to do So

The Importance of Creativity and How to Improve Yours

Building Self-Confidence

See the World Clearly

A Trilogy of Self Help Books

A New Paradigm of Truth and Happiness

Building Hope and Wonder Among Chaos

The Importance of Genius In Our World

Science, Technology, and Misc.

Future Predictions By and Engineer & Seer

The Unusual Science & Technology Bundle

Removing Limits On Our Consciousness- And Thinking Outside the Box

Universal Holistic Philosophy

Ball Lightning

Stranger Than Science Stories and Facts

Planet Earth is Conscious

Survival

Survival of Humanity Throughout the Ages

33 Incredible True Survival Stories

The Importance of Fire in History and Mythology

How to Survive Anything: From the Wilderness to Man Made Disasters

Building and Stocking a Nuclear Shelter for less than $10,000

The Human Survival Five Books Bundle

Stranger Than Science Facts and Stories

Stranger Than Science Facts and Stories Volume Two

All About Mars Journeys and Settlement

All About Mars Journeys and Settlement

All About Mars Journeys and Settlement

These books are all available in digital and printed formats from my
website and on Amazon, Barnes & Noble, Apple ITunes, and many other sites

My Books Website is: **http://mkettingtonbooks.com**

All About Mars Journeys and Settlement

Table of Contents

All About Mars Journeys and Settlement

All About Mars Journeys and Settlement

1.0 Introduction

This is a 2022 revised version of the book first published in 2020. A lot of new information has become available since then.

A manned mission to Mars has been the dream of humanity at least since the nineteenth century when we first saw details of the surface and thought there might be canals filled with water there.

Here I've looked at the history of unmanned exploration of Mars over the last fifty plus years, proposed missions to Mars, Mars Settlements, and other major issues regarding traveling to and living on Mars. Some proposals have lots of details of proposed scenarios if you want to read all of the engineering and scientific analysis work.

I grew up in the 1960s when every kid in America was fascinated with the Space program and the Astronauts. I also watched not only the Apollo 11 moon landing, but all of the successive trips to the Moon and exploration of the surface. This may be a lot of the reason I became an Engineer, worked as NASA in Houston for several years, and applied to the Astronaut Program myself.

There are some probes which reported life on Mars and then other scientists questioned the results. We are still sending unmanned probes today to try to answer those questions.

All About Mars Journeys and Settlement

2.0 Facts About Mars

The planet Mars was thought of by the ancients as the God of War. Mars is the one candidate in our Solar System which we might eventually be able to terraform to make it livable outside like Earth. You need to wear a spacesuit there now to live.

2.1 Significant Planetary Facts

Pictured in natural color in 2007

Orbital period	686.971 d (1.88082 **yr**; 668.5991 sols)
Satellites	2
Physical characteristics	
Surface area	144798500 km

	(55907000 sq mi; 0.284 Earths)		
Volume	1.6318×10^{11} km (0.151 Earths)		
Mass	6.4171×10^{23} kg (0.107 Earths)		
Mean density	3.9335 g/cm (0.1421 lb/cu in)		
Surface gravity	3.72076 m/s (12.2072 ft/s^2; 0.3794 **g**)		
Surface temp.	min	mean	max
Celsius	−143 °C	−63 °C	35 °
Fahrenheit	−226 °F	−82 °F	95 °F
Atmosphere			
Surface pressure	0.636 (0.4–0.87) kPa 0.00628 atm		
Composition by volume	95.97% carbon dioxide 1.93% argon		

1.89% nitrogen

0.146% oxygen

0.0557% carbon monoxide

0.0210% water vapor

These Facts have the following implications for a manned trip to Mars:

a) Note that Mars gravity is about one third of Earth's. This means we will need a lander and orbital launcher much more powerful than the LEM landers we used on the Moon.

b) The atmosphere of Mars is about one percent of Earths and it has water vapor in it. This means that it would be possible to use machines to absorb and separate water into hydrogen and oxygen on Mars. This would provide water and air to live as well as fuel to launch rockets.

c) Because the pressure is so low you will need to wear a spacesuit to go outside.

d) There is carbon dioxide and water ice in the planet and at the poles. These too can be harvested for water, air, and rocket fuel. Enough could be harvested to water plants in a pressurized greenhouse.

2.2 More Facts About Mars

Mars, the fourth planet from the sun, is famed for its rusty red appearance. The Red Planet is a cold, desert world with a very thin atmosphere. But the dusty, lifeless (as far as we know it) planet is far from dull.

Phenomenal dust storms can grow so large they engulf the entire planet, temperatures can get so cold that carbon dioxide in the atmosphere condenses directly into snow or frost, and marsquakes — a Mars version of an earthquake — regularly shake things up.

It, therefore, comes as no surprise that this little red rock continues to intrigue scientists and is one of the most explored bodies in the solar system, according to NASA Science.

Befitting the Red Planet's bloody color, the Romans named it after their god of war. In truth, the Romans copied the ancient Greeks, who also named the planet after their god of war, Ares.

Other civilizations also typically gave the planet names based on its color — for example, the Egyptians named it "Her Desher," meaning "the red one," while ancient Chinese astronomers dubbed it "the fire star."

WHY IS MARS CALLED THE RED PLANET?

The bright rust color Mars is known for is due to iron-rich minerals in its regolith — the loose dust and rock covering its surface. The soil of Earth is a kind of regolith, too, albeit one loaded with organic content. According to NASA, the iron minerals oxidize, or rust, causing the soil to look red.

MARS' LANDSCAPE

The planet's cold, thin atmosphere means liquid water likely cannot exist on the Martian surface for any appreciable length of time. Features called recurring slope lineae may have spurts of briny water flowing on the surface, but this evidence is disputed; some scientists argue the hydrogen spotted from orbit in this region may instead indicate briny salts. This means that although this desert planet is just half the diameter of Earth, it has the same amount of dry land.

The Red Planet is home to both the highest mountain and the deepest, longest valley in the solar system. Olympus Mons is roughly 17 miles (27 kilometers) high, about three times as tall as Mount Everest, while the Valles Marineris system of valleys — named after the Mariner 9 probe that discovered it in 1971 — reaches as deep as 6 miles (10 km) and runs east-west for roughly 2,500 miles (4,000 km), about one-fifth of the distance around Mars and close to the width of Australia.

Scientists think the Valles Marineris formed mostly by rifting of the crust as it got stretched. Individual canyons within the system are as much as 60 miles (100 km) wide. The canyons merge in the central part of the Valles Marineris in a region as much as 370 miles (600 km) wide. Large channels emerging from the ends of some canyons and layered sediments within suggest that the canyons might once have been filled with liquid water.

Mars also has the largest volcanoes in the solar system, Olympus Mons being one of them. The massive volcano, which is about 370 miles (600 km) in diameter, is wide enough to cover the state of New Mexico. Olympus Mons is a shield volcano, with slopes that rise gradually like those of Hawaiian volcanoes, and was created by eruptions of lava that flowed for long distances before solidifying. Mars also has many other kinds of volcanic landforms, from small, steep-sided cones to enormous plains coated in hardened lava. Some minor eruptions might still occur on the planet today.

All About Mars Journeys and Settlement

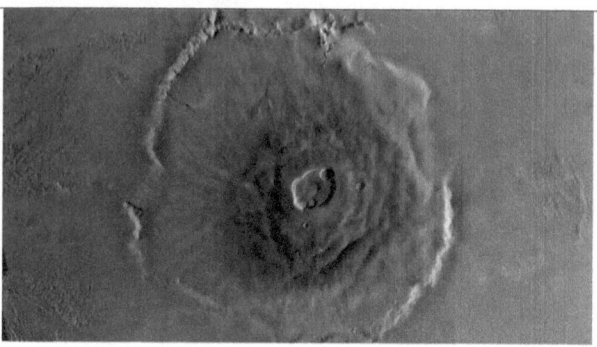

Channels, valleys and gullies are found all over Mars, and suggest that liquid water might have flowed across the planet's surface in recent times. Some channels can be 60 miles (100 km) wide and 1,200 miles (2,000 km) long. Water may still lie in cracks and pores in underground rock. A study by scientists in 2018 suggested that salty water below the Martian surface could hold a considerable amount of oxygen, which could support microbial life. However, the amount of oxygen depends on temperature and pressure; temperature changes on Mars from time to time as the tilt of its rotation axis shifts.

Many regions of Mars are flat, low-lying plains. The lowest of the northern plains are among the flattest, smoothest places in the solar system, potentially created by water that once flowed across the Martian surface. The northern hemisphere mostly lies at a lower elevation than the southern hemisphere, suggesting the crust may be thinner in the north than in the south. This difference between the north and south might be due to a very large impact shortly after the birth of Mars.

The number of craters on Mars varies dramatically from place to place, depending on how old the surface is. Much of the surface of the southern hemisphere is extremely old, and so has many craters — including the planet's largest, 1,400-mile-wide (2,300 km) Hellas Planitia — while that of northern hemisphere is younger and so has fewer craters. Some volcanoes also have just a few craters, which suggests they erupted recently, with the resulting lava covering up any old craters. Some craters have unusual-looking deposits of debris around them resembling solidified mudflows, potentially indicating that the impactor hit underground water or ice.

In 2018, the European Space Agency's Mars Express spacecraft detected what could be a slurry of water and grains underneath icy Planum Australe. (Some reports describe it as a "lake," but it's unclear how much regolith is inside the water.) This body of water is said to be about 12.4 miles (20 km) across. Its underground location is reminiscent of similar underground lakes in Antarctica, which have been found to host microbes. Late in the year, Mars Express also spied a huge, icy zone in the Red Planet's Korolev Crater.

MARS' POLAR CAPS

Vast deposits of what appear to be finely layered stacks of water ice and dust extend from the poles to latitudes of about 80 degrees in both Martian hemispheres. These were probably deposited by the atmosphere over long spans of time. On top of much of these layered deposits in both hemispheres are caps of water ice that remain frozen year-round.

Additional seasonal caps of frost appear in the wintertime. These are made of solid carbon dioxide, also known as "dry ice," which has condensed from carbon dioxide gas in the atmosphere. (Mars' think air is about 95% carbon dioxide by volume.) In the deepest part of the winter, this frost can extend from the poles to latitudes as low as 45 degrees, or halfway to the equator. The dry ice layer appears to have a fluffy texture, like freshly fallen snow, according to a report in the Journal of Geophysical Research-Planets.

MARS' CLIMATE

Mars is much colder than Earth, in large part due to its greater distance from the sun. The average temperature is about minus 80 degrees Fahrenheit (minus 60 degrees Celsius), although it can vary from minus 195 F (minus 125 C) near the poles during the winter to as much as 70 F (20 C) at midday near the equator.

The carbon-dioxide-rich atmosphere of Mars is also about 100 times less dense than Earth's on average, but it is nevertheless thick enough to support weather, clouds and winds. The density of the atmosphere varies seasonally, as winter forces carbon dioxide to freeze out of the Martian air. In the ancient past, the atmosphere was likely significantly thicker and able to support water flowing on the planet's surface. Over time, lighter molecules in the Martian atmosphere escaped under pressure from the solar wind, which affected the atmosphere because Mars does not have a global magnetic field. This process is being studied today by NASA's MAVEN (Mars Atmosphere and Volatile Evolution) mission.

NASA's Mars Reconnaissance Orbiter found the first definitive detections of carbon-dioxide snow clouds, making Mars the only body in the solar system known to host such unusual winter weather. The Red Planet also causes water-ice snow to fall from the clouds.

The dust storms on Mars are the largest in the solar system, capable of blanketing the entire Red Planet and lasting for months. One theory as to why dust storms can grow so big on Mars is because the airborne dust particles absorb sunlight, warming the Martian atmosphere in their vicinity. Warm pockets of air then flow toward colder regions, generating winds. Strong winds lift more dust off the ground, which, in turn, heats the atmosphere, raising more wind and kicking up more dust.

These dust storms can pose serious risks to robots on the Martian surface. For example, NASA's Opportunity Mars rover died after being engulfed in a giant 2018 storm, which blocked sunlight from reaching the robot's solar panels for weeks at a time.

SIZE, COMPOSITION AND STRUCTURE

Mars is 4,220 miles (6,791 km) in diameter — far smaller than Earth, which is 7,926 miles (12,756 km) wide. The Red Planet is about 10% as massive as our home world, with a gravitational pull 38% as strong. (A 100-pound person here on Earth would weigh just 62 pounds on Mars, but their mass would be the same on both planets.)

Atmospheric composition (by volume)

According to NASA, the atmosphere of Mars is 95.32% carbon dioxide, 2.7% nitrogen, 1.6% argon, 0.13% oxygen and 0.08% carbon monoxide, with minor amounts of water, nitrogen oxide, neon, hydrogen-deuterium-oxygen, krypton and xenon.

Magnetic field

Mars lost its global magnetic field about 4 billion years ago, leading to the stripping of much of its atmosphere by the solar wind. But there are regions of the planet's crust today that can be at least 10 times more strongly magnetized than anything measured on Earth, which suggests those regions are remnants of an ancient global magnetic field.

Chemical composition

Mars likely has a solid core composed of iron, nickel and sulfur. The mantle of Mars is probably similar to Earth's in that it is composed mostly of peridotite, which is made up primarily of silicon, oxygen, iron and magnesium. The crust is probably largely made of the volcanic rock basalt, which is also common in the crusts of the Earth and the moon, although some crustal rocks, especially in the northern hemisphere, may be a form of andesite, a volcanic rock that contains more silica than basalt does.

Internal structure

NASA's InSight lander has been probing the Martian interior since touching down near the planet's equator in November 2018. InSight measures and characterizes marsquakes, and mission team members are tracking wobbles in Mars' tilt over time by precisely tracking the lander's position on the planet's surface.

These data have revealed key insights about Mars' internal structure. For example, InSight team members recently estimated that the planet's core is 1,110 to 1,300 miles (1,780 to 2,080 km) wide. InSight's observations also suggest that Mars' crust is 14 to 45 miles (24 and 72 km) thick, on average, with the mantle making up the rest of the planet's (non-atmospheric) volume.

For comparison, Earth's core is about 4,400 miles (7,100 km) wide — bigger than Mars itself — and its mantle is roughly 1,800 miles (2,900 km) thick. Earth has two kinds of crust, continental and oceanic, whose average thicknesses are about 25 miles (40 km) and 5 miles (8 km), respectively.

MARS' MOONS

The two moons of Mars, Phobos and Deimos, were discovered by American astronomer Asaph Hall over the course of a week in 1877. Hall had almost given up his search for a moon of Mars, but his wife, Angelina, urged him on. He discovered Deimos the next night, and Phobos six days after that. He named the moons after the sons of the Greek war god Ares — Phobos means "fear," while Deimos means "rout."

Cratered moon Phobos

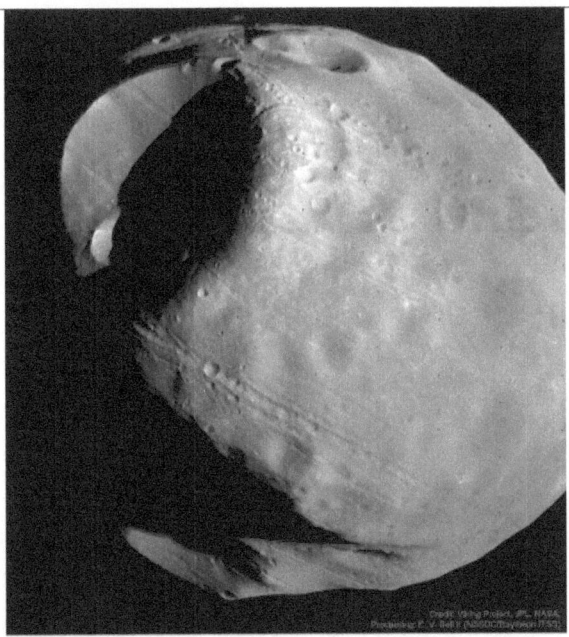

Martin Moon Phobos

Both Phobos and Deimos are apparently made of carbon-rich rock mixed with ice and are covered in dust and loose rocks. They are tiny next to Earth's moon, and are irregularly shaped, since they lack enough gravity to pull themselves into a more circular form. The widest Phobos gets is about 17 miles (27 km), and the widest Deimos gets is roughly 9 miles (15 km). (Earth's moon is 2,159 miles, or 3,475 km, wide.)

Both Mars moons are pockmarked with craters from meteor impacts. The surface of Phobos also possesses an intricate pattern of grooves, which may be cracks that formed after the impact created the moon's largest crater — a hole about 6 miles (10 km) wide, or nearly half the width of Phobos. The two Martian satellites always show the same face to their parent planet, just as our moon does to Earth.

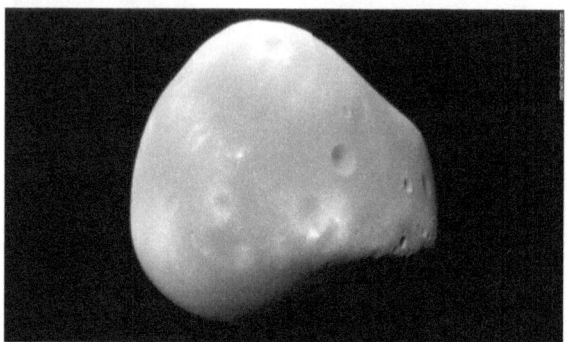

Martian Moon Deimos

It remains uncertain how Phobos and Deimos were born. They may be former asteroids that were captured by Mars' gravitational pull, or they may have formed in orbit around Mars at roughly the same time the planet came into existence. Ultraviolet light reflected from Phobos provides strong evidence that the moon is a captured asteroid ,according to astronomers at the University of Padova in Italy.

Phobos is gradually spiraling toward Mars, drawing about 6 feet (1.8 meters) closer to the Red Planet each century. Within 50 million years, Phobos will either smash into Mars or break up and form a ring of debris around the planet.

2.3 The Two year Planetary Cycle

All unmanned space probe missions and most manned mission plans to visit Mars depend on the two year orbital congruence windows between the two planets. For all long trip missions this two year window makes a lot of sense to reduce travel time and propulsion needed to a minimum.

Here is a diagram of the minimum distances between the two planets over a period of years:

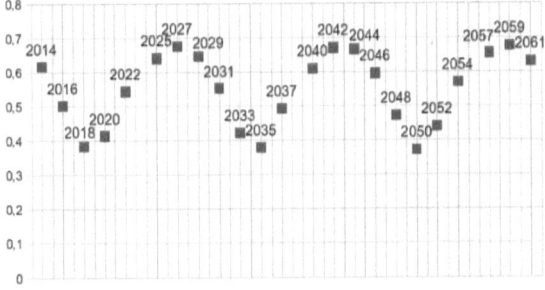

All About Mars Journeys and Settlement

3.0 Mars Unmanned Satellites and Landers
   1960s

3.1 Mars 1M spacecraft

Between 1960 and 1969, the Soviet Union launched nine probes intended to reach Mars. They all failed: three at launch; three failed to reach near-Earth orbit; one during the burn to put the spacecraft into trans-Mars trajectory; and two during the interplanetary orbit.

The Mars 1M programs (sometimes dubbed Marsnik in Western media) was the first Soviet unmanned spacecraft interplanetary exploration program, which consisted of two flyby probes launched towards Mars in October 1960, Mars 1960A and Mars 1960B (also known as Korabl 4 and Korabl 5 respectively). After launch, the third stage pumps on both launchers were unable to develop enough pressure to commence ignition, so Earth parking orbit was not achieved. The spacecraft reached an altitude of 120 km before reentry.

Mars 1962A was a Mars flyby mission, launched on October 24, 1962 and Mars 1962B an intended first Mars lander mission, launched in late December of the same year (1962). Both failed from either breaking up as they were going into Earth orbit or having the upper stage explode in orbit during the burn to put the spacecraft into trans-Mars trajectory.

Mars 1 (1962 Beta Nu 1), an automatic interplanetary spacecraft launched to Mars on November 1, 1962, was the first probe of the Soviet Mars probe program to achieve interplanetary orbit. Mars 1 was intended to fly by the planet at a distance of about 11,000 km and take images of the surface as well as send back data on cosmic radiation, micrometeoroid impacts and Mars' magnetic field, radiation environment, atmospheric structure, and possible organic compounds. Sixty-one radio transmissions were held, initially at 2-day intervals and later at 5-day intervals, from which a large amount of interplanetary data was collected. On 21 March 1963, when the spacecraft was at a distance of 106,760,000 km from Earth, on its way to Mars, communications ceased due to failure of its antenna orientation system.

In 1964, both Soviet probe launches, of Zond 1964A on June 4, and Zond 2 on November 30, (part of the Zond program), resulted in failures. Zond 1964A had a failure at launch, while communication was lost with Zond 2 en route to

Mars after a mid-course maneuver, in early May 1965.

In 1969, and as part of the Mars probe program, the Soviet Union prepared two identical 5-ton orbiters called M-69, dubbed by NASA as Mars 1969A and Mars 1969B. Both probes were lost in launch-related complications with the newly developed Proton rocket.

1970s

The USSR intended to have the first artificial satellite of Mars beating the planned American Mariner 8 and Mariner 9 Mars orbiters. In May 1971, one day after Mariner 8 malfunctioned at launch and failed to reach orbit, Cosmos 419 (Mars 1971C), a heavy probe of the Soviet Mars program M-71, also failed to launch. This spacecraft was designed as an orbiter only, while the next two probes of project M-71, Mars 2 and Mars 3, were multipurpose combinations of an orbiter and a lander with small skis-walking rovers that would be the first planet rovers outside the Moon. They were successfully launched in mid-May 1971 and reached Mars about seven months later. On November 27, 1971 the lander of Mars 2 crash-landed due to an on-board computer malfunction and became the first man-made object to reach the surface of Mars. On 2 December 1971, the Mars 3 lander became the first spacecraft to achieve a soft landing, but its transmission was interrupted after 14.5 seconds.

The Mars 2 and 3 orbiters sent back a relatively large volume of data covering the period from December 1971 to March 1972, although transmissions continued through to August. By 22 August 1972, after sending back data and a total of 60 pictures, Mars 2 and 3 concluded their missions.

The images and data enabled creation of surface relief maps, and gave information on the Martian gravity and magnetic fields.

In 1973, the Soviet Union sent four more probes to Mars: the Mars 4 and Mars 5 orbiters and the Mars 6 and Mars 7 flyby/lander combinations. All missions except Mars 7 sent back data, with Mars 5 being most successful. Mars 5 transmitted just 60 images before a loss of pressurization in the transmitter housing ended the mission. Mars 6 lander transmitted data during descent, but failed upon impact. Mars 4 flew by the planet at a range of 2200 km returning one swath of pictures and radio occultation data, which constituted the first detection of the nightside ionosphere on Mars. Mars 7 probe separated prematurely from the carrying vehicle due to a problem in the operation of one of the onboard systems (attitude control or retro-rockets) and missed the planet by 1,300 kilometers (8.7×10^{-6} au)

3.2 Mariner program

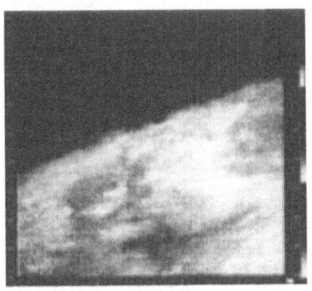

The first close-up images taken of Mars in 1965 from Mariner 4 show an area about 330 km across by 1200 km from limb to bottom of frame.

In 1964, NASA's Jet Propulsion Laboratory made two attempts at reaching Mars. Mariner 3 and Mariner 4 were identical spacecraft designed to carry out the first flybys of Mars. Mariner 3 was launched on November 5, 1964, but the shroud encasing the spacecraft atop its rocket failed to open properly, dooming the mission. Three weeks later, on November 28, 1964, Mariner 4 was launched successfully on a 7½-month voyage to Mars.

Mariner 4 flew past Mars on July 14, 1965, providing the first close-up photographs of another planet. The pictures, gradually played back to Earth from a small tape recorder on the probe, showed impact craters. It provided radically more accurate data about the planet; a surface atmospheric pressure of about 1% of Earth's and daytime temperatures of −100 °C (−148 °F) were estimated. No magnetic field or Martian radiation belts were detected. The new data meant redesigns for then planned Martian landers, and showed life would have a more difficult time surviving there than previously anticipated.

Mariner Crater, as seen by Mariner 4. The location is Phaethontis quadrangle.

NASA continued the Mariner program with another pair of Mars flyby probes, Mariner 6 and 7. They were sent at the next launch window, and reached the planet in 1969.

During the following launch window the Mariner program again suffered the loss of one of a pair of probes. Mariner 9 successfully entered orbit about Mars, the first spacecraft ever to do so, after the launch time failure of its sister ship, Mariner 8. When Mariner 9 reached Mars in 1971, it and two Soviet orbiters (Mars 2 and Mars 3, see Mars probe program above) found that a planet-wide dust storm was in progress. The mission controllers used the time spent waiting for the storm to clear to have the probe rendezvous with, and photograph, Phobos.

When the storm cleared sufficiently for Mars' surface to be photographed by Mariner 9, the pictures returned represented a substantial advance over previous missions. These pictures were the first to offer more detailed evidence that liquid water might at one time have flowed on the planetary surface. They also finally discerned the true nature of many Martian albedo features. For example, Nix Olympica was one of only a few features that could be seen during the planetary dust storm, revealing it to be the highest mountain (volcano, to be exact) on any planet in the entire Solar System, and leading to its reclassification as Olympus Mons.

3.3 Viking program

The Viking program launched the Viking 1 and Viking 2 spacecraft to Mars in 1975; the program consisted of two orbiters and two landers – these were the second and third spacecraft to successfully land on Mars.

Viking 1 lander site (1st color, July 21, 1976).

Viking 2 lander site (1st color, September 5, 1976).

The primary scientific objectives of the lander mission were to search for bio signatures and observe meteorologic, seismic and magnetic properties of Mars. The results of the biological experiments on board the Viking landers remain inconclusive, with a reanalysis of the Viking data published in 2012 suggesting signs of microbial life on Mars.

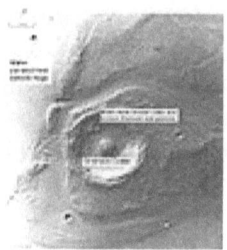

Flood erosion at Dromore crater.

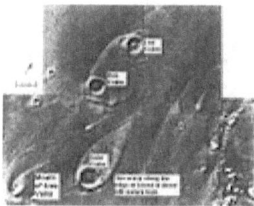

Tear-drop shaped islands at Oxia Palus.

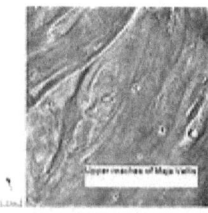

Streamlined islands in Lunae Palus.

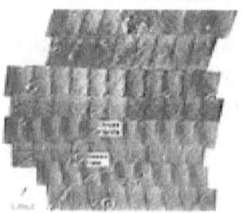

Scour patterns located in Lunae Palus.

The Viking orbiters revealed that large floods of water carved deep valleys, eroded grooves into bedrock, and traveled thousands of kilometers. Areas of branched streams, in the southern hemisphere, suggest that rain once fell.

All About Mars Journeys and Settlement

3.4 Mars Pathfinder

Sojourner takes Alpha Proton X-ray Spectrometer measurements of the Yogi Rock.

Mars Pathfinder was a U.S. spacecraft that landed a base station with a roving probe on Mars on July 4, 1997. It consisted of a lander and a small 10.6 kilograms (23 lb) wheeled robotic rover named Sojourner, which was the first rover to operate on the surface of Mars. In addition to scientific objectives, the Mars Pathfinder mission was also a "proof-of-concept" for various technologies, such as an airbag landing system and automated obstacle avoidance, both later exploited by the Mars Exploration Rovers.

All About Mars Journeys and Settlement

3.5 Mars Global Surveyor

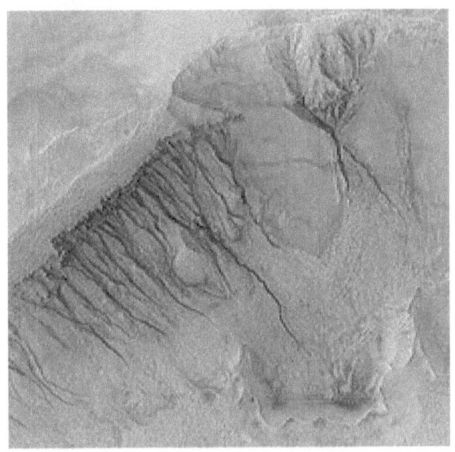

Gullies, similar to those formed on Earth, are visible on this image from Mars Global Surveyor.

After the 1992 failure of NASA's Mars Observer orbiter, NASA retooled and launched Mars Global Surveyor (MGS). Mars Global Surveyor launched on November 7, 1996, and entered orbit on September 12, 1997. After a year and a half trimming its orbit from a looping ellipse to a circular track around the planet, the spacecraft began its primary mapping mission in March 1999. It observed the planet from a low-altitude, nearly polar orbit over the course of one complete Martian year, the equivalent of nearly two Earth years. Mars Global Surveyor completed its primary mission on January 31, 2001, and completed several extended mission phases.

The mission studied the entire Martian surface, atmosphere, and interior, and returned more data about the red planet than all previous Mars missions combined. The data has been archived and remains available publicly.

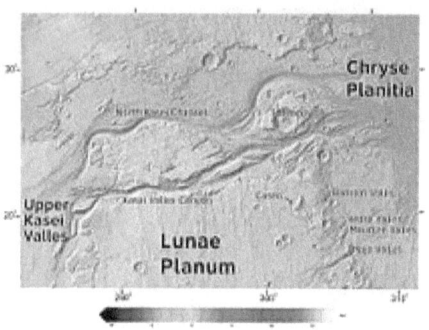

A color-coded elevation map produced from data collected by Mars Global Surveyor indicating the result of floods on Mars.

Among key scientific findings, Global Surveyor took pictures of gullies and debris flow features that suggest there may be current sources of liquid water, similar to an aquifer, at or near the surface of the planet. Similar channels on Earth are formed by flowing water, but on Mars the temperature is normally too cold and the atmosphere too thin to sustain liquid water. Nevertheless, many scientists hypothesize that liquid groundwater can sometimes surface on Mars, erode gullies and channels, and pool at the bottom before freezing and evaporating.

Magnetometer readings showed that the planet's magnetic field is not globally generated in the planet's core, but is localized in particular areas of the crust. New temperature data and close-up images of the Martian moon Phobos showed that its surface is composed of powdery material at least 1 meter (3 feet) thick, caused by millions of years of meteoroid impacts. Data from the spacecraft's laser altimeter gave scientists their first 3-D views of Mars' north polar ice cap.

Faulty software uploaded to the vehicle in June 2006 caused the spacecraft to orient its solar panels incorrectly several months later, resulting in battery overheating and subsequent failure. On November 5, 2006 MGS lost contact with Earth. NASA ended efforts to restore communication on January 28, 2007.

3.6 Mars Odyssey and Mars Express

Animation of 2001 Mars Odyssey's trajectory around Mars from 24 October 2001 to 24 October 2002

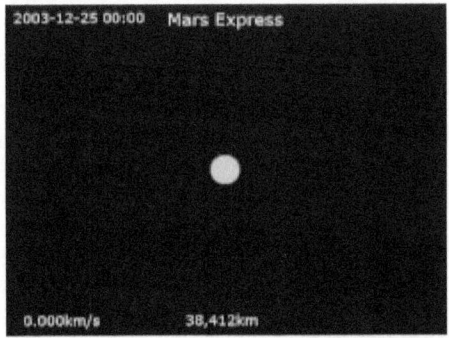

Animation of Mars Express's trajectory around Mars from 25 December 2003 to 1 January 2010

In 2001, NASA's Mars Odyssey orbiter arrived at Mars. Its mission is to use spectrometers and imagers to hunt for evidence of past or present water and volcanic activity on Mars. In

2002, it was announced that the probe's gamma-ray spectrometer and neutron spectrometer had detected large amounts of hydrogen, indicating that there are vast deposits of water ice in the upper three meters of Mars' soil within 60° latitude of the South Pole.

On June 2, 2003, the European Space Agency's Mars Express set off from Baikonur Cosmodrome to Mars. The Mars Express craft consists of the Mars Express Orbiter and the stationary lander Beagle 2. The lander carried a digging device and the smallest mass spectrometer created to date, as well as a range of other devices, on a robotic arm in order to accurately analyze soil beneath the dusty surface to look for bio signatures and biomolecules.

The orbiter entered Mars orbit on December 25, 2003, and Beagle 2 entered Mars' atmosphere the same day. However, attempts to contact the lander failed.

Communications attempts continued throughout January, but Beagle 2 was declared lost in mid-February, and a joint inquiry was launched by the UK and ESA. The Mars Express Orbiter confirmed the presence of water ice and carbon dioxide ice at the planet's South Pole, while NASA had previously confirmed their presence at the north pole of Mars.

The lander's fate remained a mystery until it was located intact on the surface of Mars in a series of images from the Mars Reconnaissance Orbiter.

The images suggest that two of the spacecraft's four solar panels failed to deploy, blocking the spacecraft's communications antenna. Beagle 2 is the first British and first European probe to achieve a soft landing on Mars.

3.7 MER and Phoenix

Polar surface as seen by the Phoenix lander.

NASA's Mars Exploration Rover Mission (MER), started in 2003, was a robotic space mission involving two rovers, Spirit (MER-A) and Opportunity, (MER-B) that explored the Martian surface geology. The mission's scientific objective was to search for and characterize a wide range of rocks and soils that hold clues to past water activity on Mars. The mission was part of NASA's Mars Exploration Program, which includes three previous successful landers: the two Viking program landers in 1976; and Mars Pathfinder probe in 1997.

All About Mars Journeys and Settlement

3.8 Mars Reconnaissance Orbiter

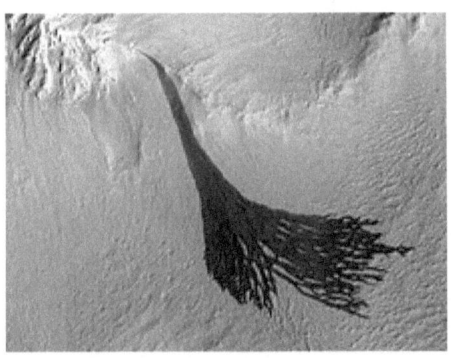

Slope streaks as seen by HiRise

The Mars Reconnaissance Orbiter (MRO) is a multipurpose spacecraft designed to conduct reconnaissance and exploration of Mars from orbit. The US $720 million spacecraft was built by Lockheed Martin under the supervision of the Jet Propulsion Laboratory, launched August 12, 2005, and entered Mars orbit on March 10, 2006.

The MRO contains a host of scientific instruments such as the HiRISE camera, CTX camera, CRISM, and SHARAD. The HiRISE camera is used to analyze Martian landforms, whereas CRISM and SHARAD can detect water, ice, and minerals on and below the surface. Additionally, MRO is paving the way for upcoming generations of spacecraft through daily monitoring of Martian weather and surface conditions, searching for future landing sites, and testing a new telecommunications system that enable it to send and receive information at an unprecedented bitrate, compared to previous Mars spacecraft. Data transfer to and from the spacecraft occurs faster than all previous interplanetary missions combined and allows it to serve as an important relay satellite for other missions.

3.9 Rosetta and Dawn swing-by's

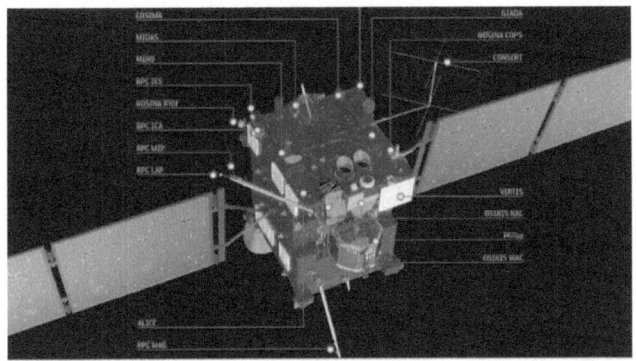

The ESA Rosetta space probe mission to the comet 67P/Churyumov-Gerasimenko flew within 250 km of Mars on February 25, 2007, in a gravitational slingshot designed to slow and redirect the spacecraft.

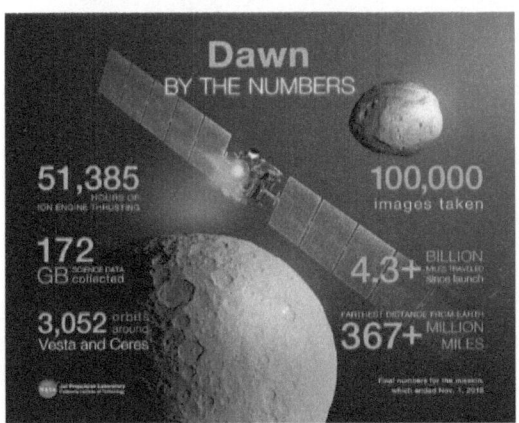

The NASA Dawn spacecraft used the gravity of Mars in 2009 to change direction and velocity on its way to Vesta, and tested out Dawn's cameras and other instruments on Mars.

All About Mars Journeys and Settlement

3.10 Fobos-Grunt

On November 8, 2011, Russia's Roscosmos launched an ambitious mission called Fobos-Grunt. It consisted of a lander aimed to retrieve a sample back to Earth from Mars' moon Phobos, and place the Chinese Yinghuo-1 probe in Mars' orbit.

The Fobos-Grunt mission suffered a complete control and communications failure shortly after launch and was left stranded in low Earth orbit, later falling back to Earth. The Yinghuo-1 satellite and Fobos-Grunt underwent destructive re-entry on January 15, 2012, finally disintegrating over the Pacific Ocean.

All About Mars Journeys and Settlement

3.11 Curiosity rover

Curiosity's view of Aeolis Mons ("Mount Sharp") foothills on August 9, 2012 EDT (white balanced image).

The NASA Mars Science Laboratory mission with its rover named Curiosity, was launched on November 26, 2011, and landed on Mars on August 6, 2012 on Aeolis Palus in Gale Crater. The rover carries instruments designed to look for past or present conditions relevant to the past or present habitability of Mars.

All About Mars Journeys and Settlement

3.12 MAVEN

NASA's MAVEN is an orbiter mission to study the upper atmosphere of Mars. It will also serve as a communications relay satellite for robotic landers and rovers on the surface of Mars. MAVEN was launched 18 November 2013 and reached Mars on 22 September 2014.

All About Mars Journeys and Settlement

3.13 Mars Orbiter Mission

The Mars Orbiter Mission, also called Mangalyaan, was launched on 5 November 2013 by the Indian Space Research Organization (ISRO). It was successfully inserted into Martian orbit on 24 September 2014. The mission is a technology demonstrator, and as secondary objective, it will also study the Martian atmosphere. This is India's first mission to Mars, and with it, ISRO became the fourth space agency to successfully reach Mars after the Soviet Union, NASA (USA) and ESA (Europe). It also made ISRO the second space agency to reach Mars orbit on its first attempt (the first national one, after the international ESA), and also the first Asian country to successfully send an orbiter to Mars. It was completed in a record low budget of $71 million, making it the least-expensive Mars mission to date.

All About Mars Journeys and Settlement

3.14 Trace Gas Orbiter and EDM

The ExoMars Trace Gas Orbiter is an atmospheric research orbiter built in collaboration between ESA and Roscosmos. It was injected into Mars orbit on 19 October 2016 to gain a better understanding of methane (CH_4) and other trace gases present in the Martian atmosphere that could be evidence for possible biological or geological activity. The Schiaparelli EDM lander was destroyed when trying to land on the surface of Mars.

All About Mars Journeys and Settlement

3.15 InSight and MarCO

In August 2012, NASA selected InSight, a $425 million lander mission with a heat flow probe and seismometer, to determine the deep interior structure of Mars.

Two   flyby CubeSats called MarCO were launched with InSight on 5 May 2018 to provide real-time telemetry during the entry and landing of InSight. The CubeSats separated from the Atlas V booster 1.5 hours after launch and traveled their own trajectories to Mars. InSight landed successfully on Mars on 26 November 2018.

All About Mars Journeys and Settlement

3.16 The Perseverence Rover

Perseverance is a NASA Mars rover that landed on 18 February 2021.

The rover will search for past life on Mars and collect soil and rock samples for future return to Earth.

Getting Perseverance's samples back to Earth will require at least two missions that are currently being planned by NASA and the European Space Agency.

What is Perseverance's mission?

Did life ever arise on Mars? For years, NASA's Mars Exploration Program has been systematically trying to find out. The agency's Spirit and Opportunity rovers showed that liquid water once existed on the surface. Building on that discovery, NASA's Curiosity rover found conditions on Mars around 4 billion years ago could have supported life as we know it. Now, Perseverance will directly search for signs of past life.

Perseverance launched on 30 July 2020 amidst the added challenge of the global COVID-19 pandemic. On 18 February 2021, it landed in Jezero crater, the site of an ancient lake and river delta. There, the rover will search for microbial fossils in rocks that formed in Mars' warm, wet past. It will also look for carbon-containing molecules called organics that form the building blocks of life on Earth. Not since 1976 has NASA directly searched for life on Mars, when the dual Viking landers performed long-shot chemistry experiments that turned up inconclusive results.

Perseverance is collecting soil and rock samples as it travels, and stores them in tubes that future missions by NASA and the European Space Agency will collect. Despite technological advances in making small, low-power science instruments for space missions, many types of laboratory analyses still can't be performed in space or can't be done very precisely. Additionally, science is all about being able to reproduce results; getting Perseverance's samples back to Earth means we could run the same science experiments in multiple laboratories.

The mission is projected to cost $2.7 billion. The Planetary Society has additional context to help you fully understand this number.

Jezero Crater on Mars

JEZERO CRATER ON MARS Jezero Crater is the landing site of NASA's Perseverance rover. In the center of this image captured by the European Space Agency's Mars Express orbiter, the remains of an ancient river delta are visible. On Earth similar deltas preserve a record of past life.

Inside Perseverance's design

Perseverance is a 1-ton, six-wheeled Mars rover the size of a compact car. Based on the same design as nuclear-powered Curiosity,

Perseverance can operate through dust storms that block sunlight required by solar-powered spacecraft. To land, Perseverance improved upon the complex seven-minute landing sequence referred to as the "seven minutes of terror." The landing involved a supersonic parachute, a thruster-powered descent, and nylon cords that lowered Perseverance the final few meters to the surface.

Using images from navigation cameras placed strategically around the rover, as well as imagery from orbital satellites such as NASA's Mars Reconnaissance Orbiter, scientists and mission operators work together to drive Perseverance to promising science areas. If a spot seems particularly interesting, Perseverance collects a sample, seals it in a small tube, and leaves the tube on the surface for return to Earth.

What instruments does Perseverance have?

At the heart of Perseverance's life-scanning tools are 2 Raman spectrometers—science instruments that shine ultraviolet light on a rock or soil patch and read the reflected light signature to determine what chemical compounds are present. Raman spectrometers are particularly well-suited to detecting organic compounds related to life as we know it. Perseverance has one on its robotic arm named SHERLOC, and another inside SuperCam, an instrument at the top of the rover's mast that also contains a laser to zap rocks several meters away!

Perseverance will look for signs of microscopic life using PIXL, a camera mounted to its robotic arm that can see features as small as a grain of salt. On Earth, we've found microscopic bacteria fossils in rocks more than 3.5 billion years old. None of Perseverance's discoveries would make sense without other science instruments to place discoveries in the proper context; particularly Mastcam-Z, two zoomable color cameras atop the rover's mast that serve as its eyes. The Planetary Society is an education and outreach partner for the instrument.

At the rover's lower-rear is RIMFAX, a radar that can detect pockets of water up to 10 meters beneath the surface. Scientists suspect that some of Mars' ancient water seeped underground, where it could still host living organisms. Also aboard Perseverance is a weather station called MEDA, as well as MOXIE, a car battery-size device that will extract carbon dioxide from Mars' atmosphere and produce oxygen, just like a tree on Earth. MOXIE will demonstrate the feasibility of manufacturing oxygen on Mars for future astronauts to breathe and use as rocket fuel.

Ingenuity, NASA's Mars helicopter

Perseverance's belly carried a small helicopter drone named Ingenuity. Once on Mars, Perseverance lowered Ingenuity to the surface and moved 100 meters away. Ever since, Ingenuity has been exploring the rover's surroundings over multiple flights. As a technology demonstration, Ingenuity is not tied to Perseverance's mission success, but we're learning important lessons about the feasibility of flying vehicles on other worlds. In 2034, NASA's Dragonfly spacecraft, an 8-bladed drone-like craft called a quadcopter, will explore Saturn's largest moon Titan.

4.0 Traveling to Mars

4.1 Months Long Journeys

Most concepts for getting to Mars involve travel in large spaceships which will take many months. Current estimates are that with a large initial boost by chemical propulsion, coasting most of the way, and slowing down, it will take six to eight months to get to Mars.

Then to do any decent type of exploration, the crew will need to remain there for another sixteen or seventeen months before the planets are aligned enough again for the journey home.

The trip to Mars aboard a traveling spaceship would be similar in many ways to being on the International Space Station for a similar period of time. The differences would be as follows:

a) The trip would be twice to three times as long when you include the rides each way and time on the surface of Mars.

b) There will be more radiation outside of Earth's orbit when travelling to Mars because Earth's magnetic field provides protection for near Earth spaceships. More protection from radiation will be required.

c) Growing your own food on the ship and on Mars will be important because taking food for everyone for four years will be prohibitively costly

d) There may need to be more of a balance of the sexes for social reasons in a multi-year mission to keep everyone mentally healthy.

e) How to keep muscles from deteriorating will be critical because we are talking about Astronauts not experiencing gravity for eight months then being asked to do strenuous activities on Mars. Likely a rotating ring to provide simulated gravity and reduce muscle deterioration will be required.

f) Communications will vary from minutes to hours when Mars and Earth are on opposite sides of the Sun. You will be posting videos like the old U.S. Mail to communicate with Earth.

4.2 Nuclear Thermal Propulsion (NTP)

Nuclear rockets could operate and boost all the time and therefore cut the travel times to and from Mars from Months to Weeks.

A nuclear thermal rocket (NTR) is a type of thermal rocket where the heat from a nuclear reaction, often nuclear fission, replaces the chemical energy of the propellants in a chemical rocket. In an NTR, a working fluid, usually liquid hydrogen, is heated to a high temperature in a nuclear reactor and then expands through a rocket nozzle to create thrust. The external nuclear heat source theoretically allows a higher effective exhaust velocity and is expected to double or triple payload capacity compared to chemical propellants that store energy internally.

NTRs have been proposed as a spacecraft propulsion technology, with the earliest ground tests occurring in 1955. The US maintained an NTR development program through 1973, when it was shut down to focus on Space Shuttle development. Although more than ten reactors of varying power output have been built and tested, as of 2019, no nuclear thermal rocket has flown.

Nuclear power in space applications that have flown include the fission-electric SNAP-10A and TOPAZ series satellites and radioisotope thermoelectric generators.

Whereas all early applications for nuclear thermal rocket propulsion used fission processes, research in the 2010s has moved to fusion approaches. The Direct Fusion Drive project at the Princeton Plasma Physics Laboratory is one such example, although "energy positive fusion has remained elusive". In 2019, the US Congress approved US $125 million in development funding for nuclear thermal propulsion rockets.

Here are six interesting facts about NTP:

1. NTP Systems Are Powered By Fission

NTP systems work by pumping a liquid propellant, most likely hydrogen, through a reactor core. Uranium atoms split apart inside the core and release heat through fission. This physical process heats up the propellant and converts it to a gas, which is expanded through a nozzle to produce thrust.

2. NTP Systems Are More Efficient Than Chemical Rockets

NTP rockets are more energy dense than chemical rockets and twice as efficient.

Engineers measure this performance as specific impulse, which is the amount of thrust you can get from a specific amount of propellant. The specific impulse of a chemical rocket that combusts liquid hydrogen and liquid oxygen is 450 seconds, exactly half the propellant efficiency of the initial target for nuclear-powered rockets (900 seconds).

This is because lighter gases are easier to accelerate. When chemical rockets are burned, they produce water vapor, a much heavier byproduct than the hydrogen that is used in a NTP system. This leads to greater efficiency and allows the rocket to travel farther on less fuel-.

3. NTP Systems Won't Be Used At Launch

NTP systems won't be used on Earth. Instead, they'll be launched into space by chemical rockets before they are turned on. NTP systems are not designed to produce the amount of thrust needed to leave the Earth's surface.

4. NTP Systems Will Provide Greater Flexibility

NTP systems offer greater flexibility for deep space missions. They can reduce travel times to Mars by up to 25% and, more importantly, limit a flight crew's exposure to cosmic radiation. They can also enable broader launch windows that are not dependent on orbital alignments and allow astronauts to abort missions and return to Earth if necessary.

5. NTP Systems Were Developed With Support From DOE

NTP is not new. It was studied by NASA and the Atomic Energy Commission (now the U.S. Department of Energy) during the 1960s as part of the Nuclear Engine for Rocket Vehicle Application program. During this time, Los Alamos National Laboratory scientists helped successfully build and test a number of nuclear rockets that current NTP designs are based off of today.

Although the program ended in 1972, research continued to improve the basic design, materials and fuels used for NTP systems.

NASA and DOE are now working with industry to develop updated nuclear thermal propulsion reactor designs. Three industry teams won a design competition in 2021 and are now further developing designs that will be submitted for evaluation for the fall of 2022.

6. NTP Systems Are Focused On Using Low-Enriched Uranium

DOE is working with NASA to help test, develop and assess the feasibility of using new fuels that require less uranium enrichment for NTP systems. This fuel may be made using new advanced manufacturing techniques and can potentially help reduce security-related costs that come with using highly enriched fuel.

Idaho National Laboratory is currently helping NASA develop and test fuel composites at its Transient Reactor Test (TREAT) facility to examine how they perform under the harsh temperatures needed for nuclear thermal propulsion. Initial testing has shown that nuclear fuels under development by NASA and DOE are capable of withstanding ramps up to operational nuclear thermal propulsion temperatures without experiencing significant damage.

All About Mars Journeys and Settlement

4.3 Mars Missions and Space Stations

A recent article on Mars Missions:

"With focus, we feel the first Mars crewed orbit mission could be accomplished by as early as 2028, with landings in follow-on missions," Tim Cichan, space exploration architect at Lockheed Martin Space, told me. "The U.S. Congress has directed NASA to study a first Mars mission by 2033."

Lockheed Martin envisions using cislunar space --- that is, the space between Earth and the Moon --- as a means for assembling and for launch of at least portions of the camp. Other crewed Mars mission scenarios advocate a more direct to Mars approach.

The orbital base camp itself would be composed of massive solar array panels for energy generation for the station's operations; a crew habitat; labs, as well as radiators, to keep both the camp's living quarters and sensitive electronics cool.

There would also be reinforced liquid oxygen and liquid hydrogen storage tanks for fuel to and from Mars and for radiation shielding for crew. And Cichan says since a reusable Mars lander isn't necessary for crew survival, it would be sent ahead to Mars uncrewed.

Base camp funding, says Cichan, would likely come from a combination of NASA, the European Space Agency (ESA), the Japanese Space Agency (JAXA) and other national space agencies. When asked about potential Mars partnerships with either SpaceX or Blue Origin, Cichan would only say, we expect NASA to have many international and commercial partners on a mission of this scale.

"We designed the Mars Base Camp architecture to fit within NASA's current human exploration budget," said Cichan. " SLS is the only launch vehicle in production today that can send Orion to deep space."

Lockheed Martin did not provide figures for such a budget. But an April 2017 report from NASA's own Office of Inspector General notes that since 2012, the SLS/Orion program has spent some $15 billion. And the office calculates that a crewed Mars mission scenario budgeted out through the 2030s would cost the space agency well over $200 billion. Such long-term budget projections seem untenable in today's Congressional fiscal environment. Thus, a Mars program of the sort that NASA and Lockheed

Martin Space envision will likely require a whole new set of unprecedented public/private partnerships.

It will also require learning new crew-rated skills in real life deep space environments.

From decades of experience with Mir, Skylab, the space shuttles (both civilian and military) and the International Space Station (ISS), we're pretty good at managing life support in planetary orbit. We're not nearly as experienced with any sort of off-world human-rated, surface-based set-up. But one of the base camps key missions would be to scope out a permanent Mars ground base.

Over the course of a 12-month period, a reusable lander operating from mars orbit could visit all five potential landing sites I noted here earlier. As a result, crews could collect data to make a better decision about where to place a semi-permanent Mars base.

All About Mars Journeys and Settlement

4.4 Concerns on the Trip to Mars

Artist's illustration of a crewed outpost on Mars.

The road to Mars is paved with peril.

Astronauts on Red Planet missions will have to contend with deep-space radiation, the effects of microgravity and the stress of confinement and isolation, all at the same time and for a long, continuous stretch. It currently takes a minimum of six months to get to Mars, after all, and just as long to get back.

And crewmembers will have to make it through this gauntlet, both physiologically and psychologically.

The spacecraft these astronauts launch aboard "will have to provide them everything they need for basic survival, but even more than that, because we expect them to be capable of doing a job — a job that has cognitive demands, a job that has physical demands," Jennifer Fogarty, the chief scientist with NASA's Human Research Program (HRP), said earlier this month during a presentation with the agency's Future In-Space Operations working group.

Many stressors

The HRP is tasked with characterizing the effects of spaceflight on astronauts and developing mitigation strategies. The program recognizes five classes of "stressor" that can significantly affect human health and performance on deep-space missions, Fogarty said. These are altered gravity fields, hostile closed environments, radiation, isolation/confinement, and distance from Earth (which means that help is very far away).

HRP scientists and other researchers around the world are trying to get a handle on all of these stressors, by performing experiments here on Earth and carefully monitoring the mental and physical health of astronauts living on the International Space Station (ISS).

The long-term goal of such work is to help enable crewed missions to Mars, which NASA wants to pull off before the end of the 2030s. Indeed, a few years ago, NASA astronaut Scott Kelly and cosmonaut Mikhail Kornienko stayed aboard the ISS for 11 months — about twice as long as the usual stint — to help researchers gauge the impact of very long space missions, such as the roundtrip journey to Mars.

It's tough to accurately characterize the toll that such a voyage will take on an astronaut, however. That's because the cumulative effect of the spaceflight stressors might be additive or synergistic, Fogarty said, and putting all of the hazards together in an experimental setting is nearly impossible.

For example, scientists perform radiation studies on lab animals here on Earth. But microgravity isn't part of that experimental picture, and adding it to the mix isn't feasible at the moment. (The ISS cannot provide deep-space radiation data, because it orbits within Earth's protective magnetosphere. And installing radiation-emitting gear aboard the orbiting lab doesn't seem like a great idea.)

Biggest concerns

Some of the stressors are more concerning than others. For example, researchers and NASA officials have repeatedly cited radiation as one of the biggest Mars-mission hazards.

High radiation exposure increases astronauts' risk of developing cancer later in life, but there are more immediate concerns as well. For instance, a recent study determined that crewmembers on a Red Planet mission will likely receive cumulative doses high enough to damage their central nervous systems. Astronauts' moods, memory and learning ability may be compromised as a result, the study found.

Fogarty mentioned another issue that requires focused research attention — spaceflight-associated neuro-ocular syndrome (SANS), also known as visual impairment/intracranial pressure (VIIP). SANS describes the potentially significant and long-lasting vision problems that spaceflight can induce in astronauts, likely because of fluid shifts that increase pressure inside the skull.

SANS "right now in low Earth orbit is very, very manageable and recoverable, but we don't know the system well enough to predict whether it will remain that way for something like an exploration mission," Fogarty said. "So, this is one of our highest-priority physiological areas that we're studying right now."

The moon as a proving ground

NASA isn't planning to go straight to Mars. The agency aims to land two astronauts near the lunar South Pole by 2024, then establish a long-term, sustainable presence on and around the moon shortly thereafter.

Indeed, the main goal of these activities, which NASA will conduct via a program called Artemis, is to learn the skills and techniques needed to send astronauts to Mars, agency officials have said.

One of Artemis' key pieces of infrastructure is a small moon-orbiting space station called the Gateway, which will serve as a hub for surface activities. For example, landers, both robotic and crewed, will descend toward the lunar surface from Gateway, and astronauts aboard the outpost will likely operate rovers from up there as well, NASA officials have said.

A great deal of research will be conducted on Gateway as well, and much of it will investigate astronauts' health and performance in a true deep-space environment. Fogarty mentioned one research strategy that may be particularly useful to planners mapping out the path to Mars — studying small samples of human tissue aboard the moon-orbiting outpost.

Such work will help researchers get around one of the biggest issues affecting studies that use rodents and other non-human animals as model organisms, Fogarty said — that of "translatability."

"How do we bridge the difference between a rat or a mouse and a human? Because it's not directly applicable, and that's plaguing terrestrial medicine and research as well," she said.

"But with the invention of, and the continued validation of, organs and tissue on a chip — those are actual human tissues, and you can connect them, and essentially you can recapitulate very sophisticated aspects of a human using these chips," Fogarty added. "I think we can make significant progress understanding the complex environment using the chip scenario as a model organism to interpret really where we're going with the human limitation."

All About Mars Journeys and Settlement

5.0 Proposals for Journeys to Mars

There are a large number of serious proposals for how man can get to Mars. This included eight major proposals in the 20th century and forty-one so far in the 21st century.

In this chapter we will try to summarize most of these proposals and cover a few of the popular ones in much more detail.

5.1 20th Century Proposals

5.1.1 Wernher von Braun proposal

Wernher von Braun was the first person to make a detailed technical study of a Mars mission. Details were published in his book Das Marsprojekt (1952, published in English as The Mars Project in 1962) and several subsequent works. Willy Ley popularized a similar mission in English in the book The Conquest of Space (1949), featuring illustrations by Chesley Bonestell. Von Braun's Mars project envisioned nearly a thousand three-stage vehicles launching from Earth to ferry parts for the Mars mission to be constructed at a space station in Earth orbit. The mission itself featured a fleet of ten spacecraft with a combined crew of 70 heading to Mars, bringing three winged surface excursion ships that would land horizontally on the surface of Mars. (Winged landing was considered possible because at the time of his proposal, the Martian atmosphere was believed to be much denser than was later found to be the case.)

In the 1956 revised vision of the Mars Project plan, published in the book The Exploration of Mars by Wernher Von Braun and Willy Ley, the size of the mission was trimmed, requiring only 400 launches to put together two ships, still carrying a winged landing vehicle. Later versions of the mission proposal, featured in the Disney "Man in Space" film series, showed nuclear-powered ion-propulsion vehicles for the interplanetary cruise.

5.1.2 U.S. proposals (1950s to 1970s)

From 1957 to 1965, work was done by General Atomics on Project Orion, a proposal for a nuclear pulse propulsion spacecraft. Orion was intended to have the ability to transport extremely large payloads compared to chemical rocketry, making crewed missions to Mars and the outer planets feasible. One of the early vehicle designs was intended to send an 800-ton payload to Mars orbit. The Partial Nuclear Test Ban Treaty of 1963 made further development unviable, and work ended in 1965.

In 1962, Aeronutronic Ford, General Dynamics and the Lockheed Missiles and Space Company made studies of Mars mission designs as part of NASA Marshall Spaceflight Center "Project EMPIRE". These studies indicated that a Mars mission (possibly including a Venus fly-by) could be done with a launch of eight Saturn V boosters and assembly in low Earth orbit, or possibly with a single launch of a hypothetical "post Saturn" heavy-lift vehicle. Although the EMPIRE missions were never proposed for funding, they were the first detailed analyses of what it would take to accomplish a human voyage to Mars using data from actual NASA spaceflight, laying the basis for future studies, including significant mission studies by TRW, North American, Philco, Lockheed, Douglas, and General Dynamics, along with several in-house NASA studies.

Following the success of the Apollo Program, von Braun advocated a crewed mission to Mars as a focus for NASA's crewed space program. Von Braun's proposal used Saturn V boosters to launch NERVA-powered upper stages that would propel two six-crew spacecraft on a dual mission in the early 1980s. The proposal was considered by President Richard Nixon but passed over in favor of the Space Shuttle.

In 1975, von Braun discussed the mission architecture that emerged from these Apollo-era studies in a recorded lecture and while doing so suggested that multiple Shuttle launches could instead be configured to lift the two Nuclear Thermal Rocket engine equipped spacecraft in smaller parts, for assembly in-orbit.

5.1.3 Soviet proposals (1956 through 1969)

The Martian Piloted Complex or "'MPK'" was a proposal by Mikhail Tikhonravov of the Soviet Union for a crewed Mars expedition, using the (then proposed) N1 rocket, in studies from 1956 to 1962. The Soviets sent many probes to Mars with some noted success stories including Mars atmospheric entry, but the overall rate of success was low. (see Mars 3)

Heavy Interplanetary Spacecraft (known by the Russian acronym TMK) was the designation of a Soviet Union space exploration proposal in the 1960s to send a crewed flight to Mars and Venus (TMK-MAVR design) without landing. The TMK spacecraft was due to launch in 1971 and make a three-year-long flight including a Mars fly-by at which time probes would have been dropped. The project was never completed because the required N1 rocket never flew successfully. The Mars Expeditionary Complex, or "'MEK'" (1969) was another Soviet proposal for a Mars expedition that would take a crew from three to six to Mars and back with a total mission duration of 630 days.

5.1.4 Case for Mars (1981–1996)

Following the Viking missions to Mars, between 1981 and 1996 a series of conferences named The Case for Mars were held at the University of Colorado at Boulder. These conferences advocated human exploration of Mars, presented concepts and technologies, and held a series of workshops to develop a baseline concept for the mission. It proposed use of in-situ resource utilization to manufacture rocket propellant for the return trip. The mission study was published in a series of proceedings volumes. Later conferences presented alternative concepts, including the "Mars Direct" concept of Robert Zubrin and David Baker; the "Footsteps to Mars" proposal of Geoffrey A. Landis, which proposed intermediate steps before the landing on Mars, including human missions to Phobos; and the "Great Exploration" proposal from Lawrence Livermore National Laboratory, among others.

In response to a presidential initiative, NASA made a study of a project for human lunar- and Mars exploration as a proposed follow-on to the International Space Station. This resulted in a report, called the 90-day study, in which the agency proposed a long-term plan consisting of completing the Space Station as "a critical next step in all our space endeavors," returning to the Moon and establishing a permanent base, and then sending astronauts to Mars. This report was widely criticized as too elaborate and expensive, and all funding for human exploration beyond Earth orbit was canceled by Congress.

5.1.5 Mars Direct (early 1990s)

Because of the greater distance, the Mars mission would be much more risky and expensive than past Moon flights. Supplies and fuel would have to be prepared for a 2-3 year round trip and the spacecraft would need at least partial shielding from ionizing radiation. A 1990 paper by Robert Zubrin and David A. Baker, then of Martin Marietta, proposed reducing the mission mass (and hence the cost) by using in situ resource utilization to manufacture propellant from the Martian Atmosphere.

This proposal drew on concepts developed by the former "Case for Mars" conference series. Over the next decade, Zubrin developed it into a mission concept, Mars Direct, which he presented in a book, The Case for Mars (1996). The mission is advocated by the Mars Society, which Zubrin founded in 1998, as practical and affordable.

5.1.6 International Space University (1991)

In 1991 in Toulouse, France, the International Space University studied an international human Mars mission. They proposed a crew of 8 traveling to Mars in a nuclear-powered vessel with artificial gravity provided by rotation. On the surface, 40 ton habitats pressurized to 10 psi (69 kPa) were powered by a 40 kW photovoltaic array.

The Mars Direct proposal includes a component for a Launch Vehicle "Ares", an Earth Return Vehicle (ERV) and a Mars Habitat Unit (MHU).

Launch Vehicle

The plan involves several launches making use of heavy-lift boosters of similar size to the Saturn V used for the Apollo missions, which would potentially be derived from Space Shuttle components. This proposed rocket is dubbed "Ares", which would use space shuttle Advanced Solid Rocket Boosters, a modified shuttle external tank, and a new Lox/LH2 third stage for the trans-Mars injection of the payload. Ares would put 121 tons into a 300 km circular orbit, and boost 47 tons toward Mars.

Earth Return Vehicle

The Earth Return Vehicle is a two-stage vehicle. The upper stage comprises the living accommodation for the crew during their six-month return trip to Earth from Mars. The lower stage contains the vehicle's rocket engines and a small chemical production plant.

Mars Habitat Unit

The Mars Habitat Unit is a 2- or 3-deck vehicle providing a comprehensive living and working environment for a Mars crew. In addition to individual sleeping quarters which provide a degree of privacy for each of the crew and a place for personal effects, the Mars Habitat Unit includes a communal living area, a small galley, exercise area, and hygiene facilities with closed-cycle water purification. The lower deck of the Mars Habitat Unit provides the primary working space for the crew: small laboratory areas for carrying out geology and life science research; storage space for samples, airlocks for reaching the surface of Mars, and a suiting-up area where crew members prepare for surface operations. Protection from harmful radiation while in space and on the surface of Mars (e.g. from solar flares) would be provided by a dedicated "storm shelter" in the core of the vehicle.

The Mars Habitat Unit would also include a small pressurized rover that is stored in the lower deck area and assembled on the surface of Mars. Powered by a methane engine, it is designed to extend the range over which astronauts can explore the surface of Mars out to 320 km.

Since it was first proposed as a part of Mars Direct, the Mars Habitat Unit has been adopted by NASA as a part of their Mars Design Reference Mission, which uses two Mars Habitat Units – one of which flies to Mars unmanned, providing a dedicated laboratory facility on Mars, together with the capacity to carry a larger rover vehicle. The second Mars Habitat Unit flies to Mars with the crew, its interior given over completely to living and storage space.

To prove the viability of the Mars Habitat Unit, the Mars Society has implemented the Mars Analogue Research Station Program (MARS), which has established a number of prototype Mars Habitat Units around the world.

5.2 21st Century Proposals

5.2.1 NASA Reference Missions (2000+)

The NASA Mars Design Reference Missions consisted of a series of conceptual design studies for human Mars missions, continued in the 21st century. Selected other US/NASA plans (1988–2009):

2000 SERT (SSP)
2001 DPT/NEXT
2002 NEP Art. Gravity
2009 DRA 5

MARPOST (2000–2005)

The Mars Piloted Orbital Station (or MARPOST) is a Russian proposed crewed orbital mission to Mars, using a nuclear reactor to run an electric rocket engine. Proposed in October 2000 as the next step for Russia in space along with participation in the International Space Station, a 30-volume draft project for MARPOST was confirmed as of 2005. Design for the ship was proposed to be ready in 2012, and the ship itself in 2021.

5.2.2 ESA Aurora program (2001+)

Artwork featuring astronauts enduring a Mars dust storm near a rover

In 2001, the European Space Agency laid out a long-term vision of sending a human mission to Mars in 2033. The project's proposed timeline would begin with robotic exploration, a proof of concept simulation of sustaining humans on Mars, and eventually a crewed mission. Objections from the participating nations of ESA and other delays have put the timeline into question, and currently ExoMars, delivered an orbiter to Mars in 2016, have come to fruition.

5.2.3 ESA/Russia plan (2002)

Another proposal for a joint ESA mission with Russia is based on two spacecraft being sent to Mars, one carrying a six-person crew and the other the expedition's supplies. The mission would take about 440 days to complete with three astronauts visiting the surface of the planet for a period of two months. The entire project would cost $20 billion and Russia would contribute 30% of these funds.

5.2.4 USA Vision for Space Exploration (2004)

Project Constellation included an Orion Mars Mission.

On 14 January 2004, George W. Bush announced the Vision for Space Exploration, an initiative of crewed space exploration. It included developing preliminary plans for a lunar outpost by 2012 and establishing an outpost by 2020. By 2005, precursor missions that would help develop the needed technology during the 2010s were tentatively outlined. On 24 September 2007, Michael Griffin, then NASA Administrator, hinted that NASA would be able to launch a human mission to Mars by 2037. The needed funds were to be generated by diverting $11 billion from space science missions to the vision for human exploration.

NASA has also discussed plans to launch Mars missions from the Moon to reduce traveling costs

5.2.5 Mars Society Germany – European Mars Mission

(EMM) (2005)

The Mars Society Germany proposed a crewed Mars mission using several launches of an improved heavy-lift version of the Ariane 5. Roughly 5 launches would be required to send a crew of 5 on a 1200 days mission, with a payload of 120,000 kg (260,000 lb). Total project was estimated to cost 10 to 15 billion Euros.

5.2.6 China National Space Administration (CNSA) (2006)

Sun Laiyan, administrator of the China National Space Administration, said on July 20, 2006 that China would start deep space exploration focusing on Mars over the next five years, during the Eleventh Five-Year Plan (2006–2010) Program period. The first uncrewed Mars exploration program could take place between 2014–2033, followed by a crewed phase in 2040–2060 in which crew members would land on Mars and return home. The Mars 500 study of 2011 prepared for this crewed mission.

5.2.7 Mars to Stay (2006)

The idea of a one-way trip to Mars has been proposed several times. In 1988, space activist Bruce Mackenzie proposed a one-way trip to Mars in a presentation at the International Space Development Conference, arguing that the mission could be done with less difficulty and expense without a return to Earth. In 2006, former NASA engineer James C. McLane III proposed a scheme to initially colonize Mars via a one-way trip by only one human. Papers discussing this concept appeared in The Space Review, Harper's Magazine, SEARCH Magazine and The New York Times.

5.2.8 NASA Design Reference Mission 5.0 (2007)

NASA released initial details of the latest version conceptual level human Mars exploration architecture in this presentation. The study further developed concepts developed in previous NASA DRM and updated it to more current launchers and technology.

5.2.9 Martian Frontier (2007–2011)

Mars 500, the longest high fidelity spaceflight simulation, ran from 2007 to 2011 in Russia and was an experiment to assess the feasibility of crewed missions to Mars.

5.2.10 NASA Design Reference Mission Architecture 5.0 (2009)

Concept for NASA's Design Reference Mission Architecture 5.0 (2009). NASA released an updated version of NASA DRM 5.0 in early 2009, featuring use of the Ares V launcher, Orion CEV, and updated mission planning in this document.

5.2.11 NASA Austere Human Missions to Mars (2009)

Extrapolated from the DRMA 5.0, plans for a crewed Mars expedition with chemical propulsion. Austere Human Missions to Mars

5.2.12 Mars orbit by the mid-2030s (2010)

In a major space policy speech at Kennedy Space Center on 15 April 2010, Barack Obama predicted a crewed Mars mission to orbit the planet by the mid-2030s, followed by a landing. This proposal was mostly supported by Congress, which approved cancelling Project Constellation in favor of a 2025 Asteroid Redirect Mission and orbiting Mars in the 2030s. The Asteroid Redirect Mission was cancelled in June 2017 and "closed out" in September of the same year.

5.2.13 Russian mission proposals (2011)

A number of Mars mission concepts and proposals have been put forth by Russian scientists. Stated dates were for a launch sometime between 2016 and 2020. The Mars probe would carry a crew of four to five cosmonauts, who would spend close to two years in space.

5.2.14 In late 2011

Russian and European space agencies successfully completed the ground-based MARS-500. The biomedical experiment simulating crewed flight to Mars was completed in Russia in July 2000.

5.2.15 2-4-2 concept (2011–2012)

In 2012, Jean-Marc Salotti published a new proposal for a crewed Mars mission. The '2-4-2' concept is based on a reduction of the crew size to 2 astronauts and the duplication of the entire mission. There are 2 astronauts in each space vehicle, there are 4 on the surface of Mars and there are 2 once again in each return vehicle. If one set of hardware runs into trouble, there are 2 astronauts ready to help the 2 others (2 for 2). This architecture simplifies the entry, descent and landing procedures by reducing the size of the landing vehicles. It also avoids the assembly of huge vehicles in LEO. The author claims that his proposal is much cheaper than the NASA reference mission without compromising the risks and can be undertaken before 2030.

5.2.16 Boeing Conceptual Space Vehicle Architecture (2012)

In 2012, a conceptual architecture was published by Boeing, United Launch Alliance, and RAL Space in Britain, laying out a possible design for a crewed Mars mission. Components of the architecture include various spacecraft for the Earth-to-Mars journey, landing, and surface stay as well as return. Some features include several uncrewed cargo landers assembled into a base on the surface of Mars. The crew would land at this base in the "Mars Personnel Lander", which could also take them back into Mars orbit. The design for the crewed interplanetary spacecraft included artificial gravity and an artificial magnetic field for radiation protection. Overall, the architecture was modular to allow for incremental R&D.

5.2.17 Mars One (2012-2019)

In 2012, a Dutch entrepreneur group began raising funds for a human Mars base to be established in 2023. The mission was intended to be primarily a one-way trip to Mars. Astronaut applications were invited from the public all over the world, for a fee.

The initial concept included an orbiter and small robotic lander in 2018, followed by a rover in 2020, and the base components in 2024. The first crew of four astronauts were to land on Mars in 2025. Then, every two years, a new crew of four would arrive. Financing was intended to come from selling the broadcasting rights of the entire training and of the flight as a reality television show, and that money would be used to contract for all hardware and launch services. In April 2015, Mars One's CEO Bas Lansdorp admitted that their 12-year plan for landing humans on Mars by 2027 is "mostly fiction". The company went bankrupt in January 2019.

5.2.18 Inspiration Mars Foundation (2013)

In 2013, the Inspiration Mars Foundation founded by Dennis Tito revealed plans of a crewed mission to fly by Mars in 2018 with support from NASA. NASA refused to fund the mission.

5.2.19 Boeing Affordable Mission (2014)

On December 2, 2014, NASA's Advanced Human Exploration Systems and Operations Mission Director Jason Crusan and Deputy Associate Administrator for Programs James Reuthner announced tentative support for the Boeing "Affordable Mars Mission Design" including radiation shielding, centrifugal artificial gravity, in-transit consumable resupply, and a lander which can return. Reuthner suggested that if adequate funding was forthcoming, the proposed mission would be expected in the early 2030s.

5.2.20 NASA Moon to Mars (2017-)

On October 8, 2015, NASA published its strategy for human exploration and colonization of Mars. The concept operates through three distinct phases leading up to fully sustained colonization.

The first stage, already underway, is the "Earth Reliant" phase. This phase continues using the International Space Station until 2024; validating deep space technologies and studying the effects of long duration space missions on the human body.

The second stage, "Proving Ground," moves away from Earth reliance and ventures into cislunar space for most of its tasks. The proposed Lunar Gateway would test deep space habitation facilities, and validate capabilities required for human exploration of Mars.

Finally, phase three is the transition to independence from Earth resources. The "Earth Independent" phase includes long term missions on the lunar surface with surface habitats that only require routine maintenance, and the harvesting of Martian resources for fuel, water, and building materials. NASA is still aiming for human missions to Mars in the 2030s, though Earth independence could take decades longer.

In November 2015, Administrator Bolden of NASA reaffirmed the goal of sending humans to Mars. He laid out 2030 as the date of a crewed surface landing, and noted that planned 2020 Mars rover would support the human mission.

In March 2019, Vice President Mike Pence declared "American Astronauts will walk on the Moon again before the end of 2024, "by any means necessary"." This reportedly prompted NASA to accelerate their plans to return to the Moon's surface by 2024. NASA says it will use the Artemis lunar program in combination with the Lunar Gateway as stepping stones to make great scientific strides "to take the next giant leap - sending astronauts to Mars".

5.2.21 Mars Base Camp (2016)

Mars Base Camp (MBC) is a US spacecraft concept that proposes to send astronauts to Mars orbit as early as 2028. The vehicle concept, developed by Lockheed Martin, would utilize both future and heritage technology as well as the Orion spacecraft built by NASA.

All About Mars Journeys and Settlement

5.3 Deep Space Transport (2017)

The 'Deep Space Transport (DST), also called Mars Transit Vehicle, is a crewed interplanetary spacecraft concept by NASA to support science exploration missions to Mars of up to 1,000 days. It would be composed of two elements: an Orion capsule and a propelled habitation module. As of April 2018, the DST is still a concept to be studied, and NASA has not officially proposed the project in an annual U.S. federal government budget cycle.

The DST vehicle would depart and return from the Lunar Gateway to be serviced and reused for a new Mars mission.

All About Mars Journeys and Settlement

5.4 NASA's Plans for Manned Missions

NASA's plans to land men on Mars as of 2020:

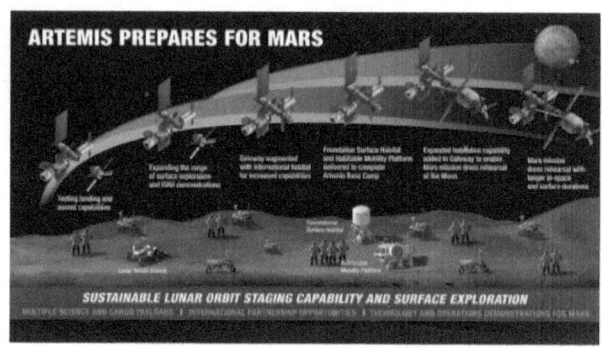

NASA's Artemis program aims to bring humans back to the Moon, with the goal of staying there for good in the interest of pursuing additional science and exploration missions, including to Mars. But how will the agency actually make it possible for people to remain on the Moon for longer-term science missions? NASA has provided some more detail about its plans with a sustainability concept it released describing some core components of the infrastructure it plans to put in place on the lunar surface.

NASA's plans focus on three key elements that would enable sustained presence and research work on the Moon's surface, including:

A lunar terrain vehicle (LTV) that would be used by crew to get around on the Moon. Essentially, this is a rover but that is piloted instead of being robotic. This wouldn't have an enclosed cockpit, so astronauts would be wearing full protective extra-vehicular activity (EVA) spacesuits while using it for short trips.

A habitable mobility platform, which would be a larger rover that is fully contained and pressurized, enabling longer trips further afield from the spacecraft landing site of up to 45 days at a time.

A lunar foundation surface habitat that could act as a more permanent, fixed location home for crew during shorter stays on the surface. This could house up to four astronauts at once, though the habitable mobility platform would be the primary active residence for surface missions, while the Gateway space station orbiting the Moon would be the main base of operations for crew not engaged in active surface exploration and science.

Like the International Space Station before it, the Gateway is designed to be scaled up over time, with new modules attached to add more crew habitation capabilities, as well as additional work and experimentation space. This will be important as it becomes the jumping off point not just for Moon surface missions, but also as a way station for exploration of Mars and beyond.

NASA also says that robotic rovers will be a key component of its Moon infrastructure, to be used for purpose including gathering data and materials for research, as well as helping to spur along the development of production of key resources for sustained presence, like water, fuel and oxygen.

The agency also includes some details about its Mars plans, including how it will send a four-person crew to the Gateway for a "multi-month stay to simulate the outbound trip to Mars." If it goes ahead as planned, this would be longest continuous human stay in deep space environs, and a key step in understanding how a human trip to Mars would work.

All About Mars Journeys and Settlement

5.5 SpaceX and Elon Musk

SpaceX has revolutionized launches into Earth orbits with reusable rockets. They deserve their own section here to review their Mars plans. SpaceX Mars transportation infrastructure (2016)

Since 2016, SpaceX publicly proposed a plan to begin the colonization of Mars by developing a high-capacity transportation infrastructure.

The ITS launch vehicle design was a large reusable booster topped by a spaceship or a tanker for in-orbit refueling. The aspirational objective is to advance the technology and infrastructure such that the first humans to Mars could potentially depart as early as 2024.

On 29 September 2017, Elon Musk announced an updated vehicle design for the Mars mission at the International Astronautical Congress. The replacement vehicle for this mission was called BFR (Big Falcon Rocket) until

2018, when it was renamed "Starship". Starship will provide the capability for on-orbit activity like satellite delivery, servicing the International Space Station, Moon missions, as well as Mars missions.

Elon Musk's TimeLine for Getting to Mars

Elon Musk has a grand plan for getting humanity out of the confines of Earth, setting off to the Moon, Mars, and even further reaches of the solar system. Musk has regularly estimated that humans could establish a city on Mars as early as 2050.

As CEO of SpaceX, he has led the development of the Starship. The rocket is designed to refuel and relaunch using liquid hydrogen and methane, unlike the rocket propellant used in the Falcon 9 and Falcon Heavy. That means astronauts will be able to set up refueling depots around the solar system, hopping from planet to planet. Still under development, the Starship could see its first commercial flight as early as 2021.

Many plans for a Mars settlement expect a community in matters of decades. The United Arab Emirates aims for a city of 600,000 by 2117. Astrobiologist Lewis Dartnell told Inverse in October that "while the first human mission to land on Mars will likely take place in the next two decades, it will probably be more like 50-100 years before substantial numbers of people have moved to Mars to live in self-sustaining towns."

SpaceX is aiming for a much, much faster timeframe, with a series of 10 launches to start a city by 2050. Here's how it looks:

SPACEX'S EVents in 2019 and 2020

The company held the first "hop tests" for its Mars-bound Starship in 2019, seeing the rocket jump a few hundred feet. SpaceX has been developing a test facility in Boca Chica, Texas, shipping over 300,000 cubic yards of locally-sourced soil. In July 2018, the firm took shipment of a 95,000-gallon liquid oxygen tank, around the same capacity as 20 tanker trucks. It's also completed a 600-kilowatt solar array and two ground station antennas that may also prove useful for Crew Dragon missions. In October 2018, it took shipment of the final major ground tank system to support the initial flights.

CEO Elon Musk previously described these tests as "fly out, turn around, accelerate back real hard and come in hot to test the heat shield because we want to have a highly reusable heat shield that's capable of absorbing the heat from interplanetary entry velocities." The firm completed its first hop test firing in April 2019, reaching a few centimeters off the ground.

Several other hops of the Starship final design we completed including a non-destructive landing of the Starship main orbital component on May 25, 2021.

SpaceX's final Starship Prototype

Assuming all goes well, it's onto the next stage. Musk claims that the first orbital launch of the full starship should occur by the end of 2022 which could help accelerate testing and move select plans to an earlier stage of the schedule.

SPACEX'S MARS PLAN: 2023

The Starship is set to embark on its first commercial flight. Jonathan Hofeller, SpaceX vice president of commercial sales, revealed at a conference in Indonesia that the plan is to host the first flight around this time.

The Starship's first voyage could see it send a commercial satellite into space for one of three telecoms firms. That sounds like a job for the Falcon 9 and Falcon Heavy, but if all goes well it could prove the Starship's viability for future missions and help fund its further development.

"You could potentially recapture a satellite and bring it down if you wanted to," Hofeller said. "It's very similar to the [space] shuttle bay in that regard. So we have this tool, and we are challenging the industry: what would you do with it?"

This could be the first year that SpaceX reaches Mars. At the International Astronautical Congress in Adelaide, Australia, in September 2017, Musk suggested this year as the point at which at least two unmanned ships could make their way to Mars. The two planets will be at an ideal point to send a rocket in 2022, a phenomenon that occurs roughly every two years.

SpaceX previously released concept art of the Starship on its way to distant planets, based around the older design rather than the more recent stainless steel iteration pictured below:

"I feel fairly confident that we can complete the ship and prepare the ship for launch in about five years," he said. "Five years feels like a long time to me."

The ships would place power, mining and life support infrastructure for future flights. They would also confirm water resources and identify hazards. Each ship would carry around 100 tons of supplies.

SPACEX'S MARS PLAN: 2024

This is the year when SpaceX is expected to send Japanese billionaire Yukazu Maezawa, alongside six to eight artists, on a trip around the moon using the Starship. While not specifically a Mars-focused mission, its success would bode well for a future manned mission. Based on Musk's February comments, this could be the first major mission for the Starship.

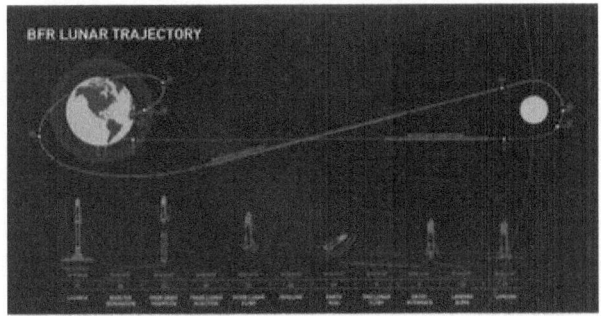

The path that Starship will take when on the Lunar Mission.

SPACEX'S MARS PLAN: 2025

It's time for another election for president of the United States. It's also the next time that the Earth and Mars are suitably aligned to send a rocket.

There's a high chance that, based on Musk's previous comments, SpaceX will not send two cargo ships to Mars in 2022 as previously suggested. If this prediction holds true, this will be the next ideal moment that SpaceX can send the cargo ships and lay the groundwork for a further mission.

If SpaceX has sent the two cargo ships by this stage, the next step will be the manned mission. The plan is to send two cargo ships, alongside two crew ships taking the first people to Mars. They will be tasked with setting up a propellant production plant, combining Martian water, ice, and carbon dioxide to create methane and liquid oxygen to fuel the ships and come back home. The humans would be tasked with collecting one ton of ice every day to fuel the plant.

The first humans will also likely have to use solar-powered hydroponics to feed the plants and grow more food. Musk said in a February interview that the technology, which allows plants to grow without soil, is already in use on Earth and the same techniques could immediately apply to the Mars colony.

A Possible Mars Base

In short, it's not going to be a leisurely visit. Musk stated at the South by Southwest Festival in Austin, Texas in March this year that Mars and the moon "are often thought of as some escape hatch for rich people, but it won't be that at all."

SPACEX'S MARS PLAN: 2026

This is the earliest point at which Musk thinks a Mars colony could take shape. The CEO has predicted a timeframe of "7 to 10 years" before the first bases take shape.

This will expand on the work left behind by the first humans. Paul Wooster, Principal Mars development engineer for SpaceX, explained that "the idea would be to expand out, start off not just with an outpost, but grow into a larger base, not just like there are in Antarctica, but really a village, a town, growing into a city and then multiple cities on Mars." The larger cities would offer habitats, greenhouses, life support, and enable new experiments that help to answer some of the big questions about life on Mars.

A potential future Mars city.

SPACEX'S MARS PLAN: 2026

This could be the next time that SpaceX sends more ships to Mars. Musk explained on Twitter that the company could use 10 orbital synchronizations to complete a city by the year 2050. With the two planets set to align in February 2027, this could be about the right time to complete another launch.

SPACEX'S MARS PLAN: BEYOND

By the end of the next decade, SpaceX expects to have some sort of settlement on Mars. Musk has said there's a 70 percent chance he'll visit Mars himself in his lifetime, perhaps paying a visit to this developing colony. That is, depending on how the first settlements go — Musk said in 2016 that "probably people will die," but "ultimately, it will be very safe to go to Mars, and it will be very comfortable."

Mars could perhaps serve as a base for more ambitious missions, with Musk describing the Starship as "really intended as an interplanetary transport system that's capable of getting from Earth to anywhere in the solar system as you establish propellant depots along the way."

Beyond transforming humanity into a space-faring civilization, it could also preserve the species. SpaceX president Gwynne Shotwell said in April that "if something were to happen on Earth, you need humans living somewhere else...I think you need multiple paths to survival, and this is one of them."

6.0 Mars Settlements
Some general ideas for Mars settlements include the following as well as other options like living under a Martian lake.

Equatorial regions

Mars Odyssey found what appear to be natural caves near the volcano Arsia Mons. It has been speculated that settlers could benefit from the shelter that these or similar structures could provide from radiation and micrometeoroids. Geothermal energy is also suspected in the equatorial regions.

Lava tubes

Several possible Martian lava tube skylights have been located on the flanks of Arsia Mons. Earth based examples indicate that some should have lengthy passages offering complete protection from radiation and be relatively easy to seal using on-site materials, especially in small subsections.

Hellas Planitia

Hellas Planitia is the lowest lying plain below the Martian geodetic datum. The air pressure is relatively higher in this place when compared to the rest of Mars.

We will explore some of these ideas in more detail in this chapter.

6.1 Creative Settlement Designs

On the Surface these Stunning Designs Show What Our Future on Mars Might Look Like

Max Rymsha, from the Ukraine, was one of the winners in the HP Mars Home Planet Rendering Challenge. They won for their piece "Between the Red Mountains" as part of the Rendering challenge.

A recent contest challenged participants to create utopian designs of future human Mars settlements, and their creations are stunning.

In the HP Mars Home Planet Rendering Challenge, over 87,000 people from all over the world flexed their creative muscles to design the perfect colony on the Red Planet. Last summer, when HP launched the challenge, the participants started working on their designs, and the winners were announced on Aug. 14.

This challenge wasn't just about creating a pretty, futuristic-looking, idealistic Martian colony. Indeed, the designs also had to show how the settlements would support 1 million colonists.

The surface of the Red Planet is harsh, with an extremely thin atmosphere, intense radiation and dust storms that occasionally envelop the planet. The participants' designs were judged on originality, creativity, rendering quality and Mars physics (or how the design would realistically work on the actual Martian surface), HP included in a statement. The designs must take into account atmospheric conditions, gravity, the soil, the surface terrain, radiation, drinking water, and air, the statement added.

6.2 The Homestead Project

The Homestead Project: Making a Mars Settlement a Reality. This concept provides a clear vision for early life on the Martian frontier. It is built primarily from local materials, providing early life support and industrial capabilities, and will begin full-scale settlement of the Red Planet.

The Mars Foundation's hope for humanity's future on Mars is neatly summed up by their slogan: "To arrive, survive and thrive!"

In July at the International Conference on Environmental Systems (SAE-ICES) in Rome, the group presented plans for a permanent settlement they believe can be built using near-term technologies and resources already available on Mars.

She's a brick house

Designs call for large masonry arches and vaulted ceilings and domed skylights built with bricks baked from Martian soil and stones cut from Martian quarries.

Bruce Mackenzie, aco-founder of the group and a former member of the National Space Society's board of directors, has been preaching the benefits of brick as an ideal building material for a Martian settlement for years.

"There are a number of ways you can make it, including just scooping up the soil, putting it in a mold, and compressing and heating it," he said. "You can also melt it and make glass, and it can be glued together."

Brick is also easy to manufacture, Mackenzie said, and quality control for brick is not critical the way it is for other materials like fiberglass.

Additional materials—such as steel, aluminum, ceramic, glass and plastics--will also be needed for the settlement's construction but the group believes these materials can be manufactured using local Martian resources.

"The industry and the technology that you need to produce these materials we'll have on hand, "said Joseph Palaia, an MIT nuclear engineering graduate student involved in the settlement design. "It's based on last century's industrial engineering technology."

Compared to the cramped quarters within space shuttles and the International Space Station, the Martian settlement will be large-- approximately 27,000 square feet--and will initially house a dozen settlers.

"We're not putting them in a trailer somewhere," said Mark Homnick, another Mars Foundation co-founder and a retired engineer who designed wafer-fabrication plants for Intel. "This thing is roomy and intended for permanent habitation."

As more settlers arrive, the site will be expanded and will ultimately be able to accommodate approximately 100 people, the group said.

The settlement will be contained within an artificial atmosphere and pressurized using gases found on Mars like carbon, nitrogen and argon, the group said. Oxygen will be stripped from water molecules using electrolysis and will also added to the mix.

Ideal conditions?

Conditions on Mars, however, are not exactly colonization-friendly and compared to Earth, in fact, they can seem downright hostile. Morning temperatures on the desert planet can dip can below -76 degrees Fahrenheit (-60 Celsius) and enormous dust storms sweep across its barren rocky fields at speeds of over 60 miles per hour.

A wispy atmosphere, combined with the lack of a planetary magnetic field, means that the air pressure on Mars is only a tiny fraction of Earth's and that harmful radiation from solar winds, cosmic rays and solar flares routinely bombard its surface. Factor in a minimum 6-month commute and a communications delay that can reach over 40-minutes and an obvious question arises: Why would anyone want to go to Mars? Let alone live there?

One reason, said Palaia, is because it's there. "We will go to Mars for the challenge," he said. "Anything short of Martian settlement will be too easy an undertaking."

Mars is also scientifically interesting-- geologically and perhaps even biologically--and research conducted from a permanent base would be more efficient and less costly, the group said.

Compared to a round-trip exploratory mission, the group believes a permanent settlement may also be safer. Broken parts, for example, could be manufactured and replaced on-site, eliminating the need to haul heavy spare parts or risks dangerous shipping delays.

"Anything that is high-mass and low tech, we're going to make there on Mars," said Palaia."Anything that is really high tech--like sensors, motors and complex mechanism--most of those things are relatively low mass and can be imported from Earth."

The group recommends sending a minimum amount of resources to Mars beforehand, a process known as bootstrapping. When the settlers arrive on Mars, they can use the prepared materials, along with local resources, to construct the settlement.

One possible scenario, the group proposes is to send small gas tanks ahead that store methane and oxygen extracted from the atmosphere. When the settlers arrive, they can then use that equipment and stored gas to build things like steel production plants.

Finally, Mars will be an integral part of an inter-solar system economy that the group believes will develop within the next century, one based on the convergence of four frontiers: Earth, the Moon, asteroids, and Mars--including its own rocky satellites, Phobos and Deimos.

Mars will catalyze the development of the other frontiers, said Homnick, acting as a supply house for vital resources like nitrogen, carbon dioxide and water for the moon and asteroids, places where such things are scarce or nonexistent.

Many of the technologies developed for use on Mars will also have applications for the other frontiers, the group said. For example, life support systems and mining equipment developed for use on Mars could also be used on the moon.

The group strongly supports President Bush's Moon, Mars and Beyond vision and said they are not trying to compete with NASA or any other space organization.

"We kind of look at NASA and the European Space Agency as analogous to Lewis and Clark in the old west," Homnick said. "They blaze the trail, go out to explore and do the science. Well, we are analogous to the pioneers--we follow the trail that they blazed, and we make the new frontier home and we add value."

Instead, the group believes that different agencies can benefit from one another and the colonization of space can be sped up.

"We hope they succeed because they'll help us succeed," said Palaia. It's all about location.

While drawing up plans for the settlement, the group restricted themselves to existing--or extrapolations of existing--technologies. Despite this limitation, the group believes the first stages of a Martian settlement could be in place as soon as 2025.

After studying Martian survey data collected by NASA, the group chose Candor Chasma as a tentative site for the settlement. Candor Chasma is a group of mesas located within an enormous canyon system on Mars known as the Valles Marineris.

In addition to being geologically varied and scientifically interesting, Candor Chasma is also relatively flat and situated near the planet's equator, factors that are important for shuttle take offs and landings.

The settlement will be an oasis built for posterity, one the group believes future generations will come to regard as "a place of veneration and pilgrimage."

With this in mind, the group's settlement designs call for the planting of a First Tree. The tree—the species of which will be determined later--will be planted in front of the settlement's main entrance and its seeds will be transplanted to new parts of the settlement as it expands.

"That was very important to us," said Palaia. "We wanted to have this in there as a symbol of bringing life to [Mars]."

Mackenzie and Homnick are both middle-aged and doubt they'll be able to go to Mars themselves. But Palaila, 25, thinks he may have a chance.

"It's been my life obsession since I was very young," he said. Whether he'll be able to remain on Mars permanently, however, is another matter. "It's a point of contention with my wife," he said.

All About Mars Journeys and Settlement

All About Mars Journeys and Settlement

6.3 Sub lake Settlements

Terraforming a world is a breathtaking task, one often thought about in relation to making Mars into a benign environment for human settlers. But there are less challenging alternatives for providing shelter to sustain a colony. As Robert Zubrin explains in the essay below, ice-covered lakes are an option that can offer needed resources while protecting colonists from radiation. The founder of the Mars Society and author of several books and numerous papers, Zubrin is the originator of the Mars Direct concept, which envisions exploration using current and near-term technologies. We've examined many of his ideas on interstellar flight, including magsail braking and the nuclear salt water rocket concept, in these pages. Now president of Pioneer Astronautics, Zubrin's latest book is The Case for Space: How the Revolution in Spaceflight Opens Up a Future of Limitless Possibility, recently published by Prometheus Books. By Robert Zubrin

Abstract

This paper examines the possibilities of establishing Martian settlements beneath the surface of ice-covered lakes. It is shown that such settlements offer many advantages, including the ability to rapidly engineer very large volumes of pressurized space, comprehensive radiation protection, highly efficient power generation, temperature regulation, copious resource availability, outdoor recreation, and the creation of a vibrant local biosphere supporting both the nutritional and aesthetic needs of a growing human population.

Introduction

The surface of Mars offers many challenges to human settlement. Atmospheric pressure is only about 1 percent that of Earth, imposing a necessity for pressurized habits, making spacesuits necessary for outdoor activity, and providing less than optimum shielding against cosmic radiation. For these reasons some have proposed creating large subsurface structures, comparable to city subway systems, to provide pressurized well-shielded volumes for human habitation. The civil engineering challenges of constructing such systems, however, are quite formidable.

Moreover, food for such settlements would have to be grown in greenhouses, limiting potential acreage, and imposing either huge power requirements if placed underground, or the necessity of building large transparent pressurized structures on the surface. Water is available on the Martian surface as either ice or permafrost. These materials can be mined and the product transported to the base, but the logistics of doing so, while greatly superior to anything possible on the Moon, are considerably less convenient than the direct access to liquid water available to nearly all human settlements on Earth. While daytime temperatures are acceptably close to 0 C, nighttime temperatures drop to -90 C, imposing issues on machinery and surface greenhouses. Yet despite the cold night temperatures, the efficiency of nuclear power is impaired by the necessity of rejecting waste heat to a near-vacuum environment.

All of these difficulties could readily be solved by terraforming the planet. However, that is an enormous project whose vast scale will require an already-existing Martian civilization of considerable size and industrial power to be seriously undertaken. For this reason, some have proposed the idea of "para terraforming," that is, roofing over a more limited region of the Red Planet, such as the Valles Marineris, and terraforming just that part. But building such a roof would itself be a much larger engineering project than any yet done in human history.

There are, however, locations on Mars that have already been roofed over. These are the planet's numerous ice-filled craters.

Making Lakes on Mars

Earth's Arctic and Antarctic regions feature numerous permanently ice covered or "sub glacial" lakes. These lakes have been shown to support active microbial and planktonic ecosystems.

Most sub-Arctic and high latitude temperate lakes are ice-covered in winter, but many members of their aquatic communities remain highly active, a fact well-known to ice fishermen.

Could there be comparable ice-covered lakes on Mars?

At the moment, it appears that there are not. The ESA Mars Express orbiter has detected highly-saline liquid water deep underground on Mars using ground penetrating radar, and such environments are of great interest for scientific sampling via drilling. But to be of use for settlement, we need ice-covered lakes that are directly accessible from the surface. There are plenty of ice-filled craters on Mars. These are not lakes, however, as while composed of nearly pure water ice, they are frozen top to bottom. But might this shortcoming be correctable?

I believe so. Let us examine the problem by considering an example.

Korolev is an ice-filled impact crater in the Mare Boreum quadrangle of Mars, located at 73° north latitude and 165° east longitude. It is 81.4 kilometers in diameter and contains about 2,200 cubic kilometers of water ice, similar in volume to Great Bear Lake in northern Canada. Why not use a nuclear reactor to melt the water under the ice to create a huge ice-covered lake?

Korolev Crater could provide a home for sub lake city on Mars. Photo by ESA/DLR.

Let's do the math. Melting ice at 0 C requires 334 kJ/kg. We will need to supply this plus another 200 kJ/kg, assuming that the ice's initial temperature is -100 C, for 534 kJ/kg in all. Ice has a density of 0.92 kg/liter, so melting 1 cubic kilometer of ice would require 4.9 x 1017 J, or 15.6 GW-years of energy. A 1 GWe nuclear power plant on Earth requires about 3 GWt of thermal power generation. This would also be true in the case of a power plant located adjacent to Korolev, since it would be using the ice water it was creating in the crater as an excellent heat rejection medium. With the aid of 5 such installations, using both their waste heat and the dissipation from their electric power systems, we could melt a cubic kilometer of ice every year.

Korolev averages 500 m in depth, which is much deeper than we need. So rather than try to melt it all the way through, an optimized strategy might be to focus on coastal regions with an average depth of perhaps 40 meters. In that case, each cubic kilometer of ice melted would open 25 square kilometers of liquid lake for settlement. Alternatively, we could just choose a smaller crater with less depth, and melt the whole thing, except the ice cover at its top.

Housing in a Martian Lake

On Earth, 10 meters of water creates one atmosphere of pressure. Because Martian gravity is only 38 percent as great as that of Earth, 26 meters of water would be required to create the same pressure. But so much pressure is not necessary. With as little as 10 meters of water above, we would still have 0.38 bar of outside pressure, or 5.6 psi, allowing a 3 psi oxygen/2.6 psi nitrogen atmosphere comparable to that used on the Skylab space station. Reducing nitrogen atmospheric content in this way could also be advantageous because nitrogen is only a small minority constituent of the Martian atmosphere, making it harder to come by on Mars, and limiting the nitrogen fraction of breathing air would also facilitate traveling to lower pressure environments without fear of getting the bends. Ten meters of water above an underwater habitat would also provide shielding against cosmic rays equivalent to that provided by Earth's atmosphere at sea level.

Construction of the habitats could be done using any of the methods employed for underwater habitats on Earth. These include closed pressure vessels, like submarines, or open-bottom systems, like diving bells. The latter offer the advantage of minimizing structural mass since they have an interior pressure nearly equal to that of the surrounding environment, and direct easy access to the sea via their bottom doors, without any need for airlocks. Thus, while closed submarines are probably better for travel, as their occupants do not experience pressure changes with depth, open bottom habitats offer superior options for settlement. We will therefore focus our interest on the latter.

Consider an open-bottom settlement module consisting of a dome 100 m in diameter, whose peak is 4 meters below the surface and whose base is 16 meters below the surface. The dome thus has four decks, with 3 meters of head space for each. The dome is in tension, because all the air in it is all at a pressure of 9 psi, corresponding to the lake water pressure at its base, while the lake water pressure at its top is only about 2.2 psi, for an outward pressure on the dome material near the top of 6.8 psi. The dome has a radius of curvature of 110 m.

The required yield stress of the material composing a pressurized sphere is given by:

$\sigma = xPR/2t$

Where σ is the yield stress, P is the pressure, R is the radius, t is the dome thickness, and x is the safety factor. Let's say the dome is made of steel with a yield stress of 100,000 psi and x=2. In that case, equation (1) says that:

100,000 = (6.8)(110)/t, or t= 0.0075 m = 7.5 mm.

The mass of the steel would be about 600 tons. That's not too bad, for creating a habitat with about 30,000 square meters of living space.

If instead of using steel, we made a tent dome from spectra fabric, which has 4 times the strength of steel and 1/9th the density, the mass of the dome would only need to be about 17 tons. It would, however, need to be tied down around its circumference. Ballast weights of 90,000 tons of rocks could be used for this purpose. Otherwise the tie down lines could be anchored to stakes driven deep into the frozen ground under the lake.

An attractive alternative to these engineering methods for creating a dome out of manufactured materials could be to simply melt the dome out of the ice covering the lake itself. For example, let's say the ice cover is 20 m thick, and we melt a dome into it that is 12 m tall, 100 m in diameter, and has a radius of curvature of 110 m. Filling this with an oxygen/nitrogen gas mixture would provide a habitat of equal size to that discussed above. The pressure under 20 m of ice (density = 0.92) is 0.7 bar, or 10.3 psi. The roof of the dome is under 8 m of ice, whose mass exerts of compressive pressure of 0.28 bar, or 4.1 psi, leaving a pressure difference of 6.2 psi to be held by the strength of the ice. The tensile strength of ice is about 150 psi, so sticking these values into equation (1) we find that the safety factor, x, at the dome's thinnest point would be:

$$150 = x(6.2)(110)/[(8)(2)], \text{ or } x = 3.52$$

This safety factor is more than adequate. Networks of domes of this size could be melted into the ice cover, linked by tunnels through the thick material at their bases. If domes with a much larger radius of curvature were desired, the ice could be greatly strengthened by freezing a spectra net into it.

The mass of ice melted to create each such dome is about 80,000 tons, requiring 1 MWt-year of energy to do the melting. It would also require about 90 tons of oxygen to fill the dome with gas. This could be generated via water electrolysis. Assuming 80% efficient electrolysis units, this would require 1950 GJ, or 62 kWe-year of electric power to produce. Such large habitation domes could therefore be constructed and filled with breathable gas well in advance of the creation of the lake using much more modest power sources.

Compressive habitation structures can be created under ice that are much larger still. This is so because ice has 92 percent the density of water, so that if a 50 meters deep column of ice beneath the lake's ice surface were melted, it would yield a column of water 42 meters deep and 8 meters of void, which could be filled with air.

So, let's say we had an ice crater, section of an ice crater, or even a glacier 5 km in radius and 70 meters or more deep. We melt a section of it starting 20 m under the top of the ice and going down 50 m. As noted, this would create a headroom space 4 m thick above the water. The ice above this void would have a weight of 7 psi, so we would fill the void with an oxygen/nitrogen gas mixture with a pressure of 6.999 psi. This would negate almost all the weight to leave the ice roof in an extremely mild state of compression. (Mild compression is preferred to mild tension, because the compressive strength of ice is about 1500 psi – ten times the tensile strength.) Under such circumstances the radius of curvature of the overhanging surface could be unlimited. As a result, a pressurized and amply shielded habitable region of 78 square kilometers would be created. Habitats could be placed on rafts or houseboats on this indoor lake, or an ice shelf formed to provide a solid floor for conventional buildings over much of it.

The total amount of water that would need to be melted to create this indoor lake city would be 4 cubic kilometers. This could be done in about 4 years by our proposed 5 GWe power system. Further heating would continue to expand the habitable region laterally over time. If the lake were deep, so that there was ice beneath the water column, it would gradually melt, increasing the headroom over the settlement as well.

Terraforming the Lake

The living environment of the sub lake Mars settlement need not be limited to the interior of the air-filled habitats. By melting the ice, we are creating the potential for a vibrant surrounding aquatic biosphere, which could be readily visited by Mars colonists wearing ordinary wet suits and SCUBA gear.

The lake is being melted using hot water produced by the heat rejection of onshore or floating nuclear reactors. If the heat is rejected near the bottom of the lake, forceful upwelling will occur, powerfully fertilizing the lake water with mineral nutrients.

Assuming that the ice cover is reduced to less than 30 meters, there will be enough natural light during daytime to support phytoplankton growth, as has been observed in the Earth's Arctic Ocean. The lake's primary biological productivity could be greatly augmented, however, by the addition of artificial light.

The Arctic Ocean exhibits high biological activity as far north as 75 N, where the sea receives an average day/night year-round solar illumination of about 50 $W/m2$. If we take this as our standard, then each GW of our available electric power could be used to illuminate 20 square kilometers of lake. Combined with the mineral-rich water produced by thermal upwelling, and artificial delivery of CO_2 from the Martian atmosphere as required, this illumination could serve to create an extremely productive biosphere in the waters surrounding the settlement.

The first organisms to be released into the lake should be photosynthetic phytoplankton and other algae, including macroscopic forms such as kelp. These would serve to oxygenate the water. Once that is done, animals could be released, starting with zooplankton, with a wide range of aquatic macrofauna, potentially including sponges, corals, worms, mollusks, arthropods, and fish coming next. Penguins and sea otters could follow.

As the lake continues to grow, its cities would multiply, giving birth to a new branch of human civilization, supported by and supporting a lively new biosphere on a new world.

Conclusion

We find that the best places to settle Mars could be under water. By creating lakes beneath the surface of ice-covered craters, we can create miniature worlds, providing acceptable pressure, temperature, radiation protection, voluminous living space, and everything else needed for life and civilization. The sub lake cities of Mars could serve as bases for the exploration and development of the Red Planet, providing homes within which new nations can be born and grow in size, technological ability, and industrial capacity, until such time as they can wield sufficient power to go forth and take on the challenge of terraforming Mars itself. Technical processes that have accumulated as much mineral ores as we could wish for.

6.4 Mars Colony in a Lava Tube

This was a project of the NASA Institute of Advanced Projects. The final report was in 2004.

The final report is divided into 10 parts:

- Project Summary
- Introduction
- Enabling Technologies Identification
- Essential Tasks Identification
- Demonstration Missions
- Technology Trials
- Planetary Protection Protocol Development
- Education and Outreach
- Conclusions

Section 1, the Project summary summarizes the entire project and claims that "This project developed a revolutionary system to exploit the novel idea of extraterrestrial cave use" and explaining that two experiments or "Missions" were tested to gather data.

Section 2, the Introduction answers the question of "why caves [for Martian research bases]?" and provides a variety of different answers to the advantages of using caves as a foothold in Martian exploration such as:

It is surmised that atmosphere temperature variations are less experienced in caves than on the surface of Mars.

Caves protect lifeforms and equipment from harmful solar and galactic radiation. Caves protect from dust storms and micrometeorite impacts.

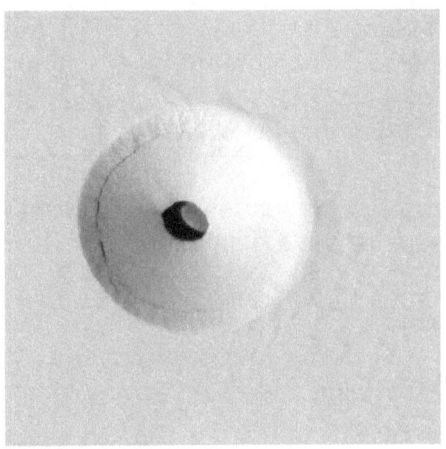

A Mars Lava Tube

Exploring caves is a key scientific interest as it makes it easy to study the geology, history, and possible presence of life on Mars without heavy excavation equipment.

The ability of pressurizing caves to make them more habitable to human lifeforms.

Allows the easy extraction of subsurface materials such as ice and minerals.

This section also contains some speculation on the existence and locations of such caves and what types of caves exist on Mars however it is largely outdated by newer research such as the HiRISE and THEMIS missions.

Section 3, a description of the Enabling Technologies analyses a number of Innovations necessary for the utilization and assigns them a Technology Readiness Level. For example, the innovation "Foamed-in-place Airlocks" are assigned a TRL of 5, while the "Inert Pressurization of Caves" is assigned a TRL of 2.

Section 4 describes the "Essential Tasks" necessary for cave habitation.

These are:
- Finding extraterrestrial caves
- Protecting the scientific environment inside of a cave
- Dealing with the dark (providing lighting solutions for the interior of the habitats)

Cave Life support

The publication discusses each of these topics in detail and highlights the novel idea of using luminescent bacteria as a lighting backup solution and suggests lighting the habitat using "light piping" technology. The article also discusses skylights and radiation proof glass at length however this is probably due to the lack of advanced solar panels and LED lighting technology available during the publication in 2000.

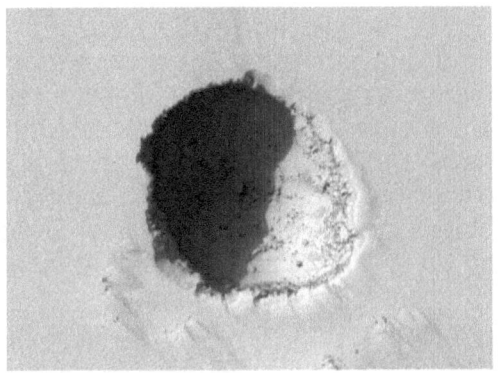

Another Mars Lava Cave

Section 5 contains information on the "Demonstration Missions", specifically the "Mouse Mission to Inner Space" (MOMIS) and the Human equivalent, "HUMIS". The idea was to develop preliminary versions of some aspects of a Mars cave habitat such as using argon breathing mixtures and other new life support systems on mouse test subjects. The MOMIS experiment has successfully completed multiple runs however the HUMIS experiment was deemed out of the scope of the investigation and although efforts were made to find test sites, the work done was reflected in a "Cave Astrobiology" exploration-level class at Penn State College during the spring semester of 2004.

Section 6 covers the different "Technology Trials" performed. First, Inflatable habitats were investigated to provide a "shirtsleeve indoor environment" for the astronauts. The article further suggests that if the cave's cross sectional surface area is properly sized, an inflatable cave liner could be placed in the cave and inflated requiring no additional support systems. The article then suggests using a dual-liner system in which an outer liner provides a surface against the cave surface and a pressure seal and an inner liner provides a habitat for the astronauts. Machinery and life support systems could be placed in between the redundant liners. The report also outlines methods of folding, manufacturing, transporting, replacing, and inflating these liners. Another main topic of this section is the "foamed in place" airlocks. These are designed to be shape-conforming to highly irregular openings along with easy to deploy and leak tight. Their final proposed system is an airlock unit with multiple extending telescoping legs to all of the cave walls. The space between the airlocks and the cave walls are then filled with hardening, spray-able, airtight foam.

Next, the report outlines methods by which an inert pressure atmosphere could be created by pressurizing the gasses present on Mars, particularly Argon. This would allow human scientists only to wear breathing apparatuses and not require full pressure suits. It is suggested that cavernous spaces not be filled with oxygen or other reactive gasses as this would nullify any potential scientific value of the cave along with potentially being harmful to the humans breathing in the atmosphere inside. Finally, this section covers a system that would allow communication networks inside caves. This was also tested in a real cave (Robertson's Cave) and future modifications are suggested for increasing bandwidth and signal strength.

Section 7 covers the development of a Planetary Protection Protocol and highlights its importance when exploring Martian caves and suggests using "sterilized micro-robots" to perform exploration and science.

Section 8 is named "Education and Outreach". It contains information on the spinoff science-fair experiments generated from this report and the other outreach impacts that this report and creating it had. This section also outlines educational activities for schools such as a "Find the lava tube activity" and "The Mousetronauts Program".

Sections 9 and 10 conclude the report and cite references for further reading.

6.5 Plans for First Manned Landings

Due to the distance to get to Mars and its one third gravity it's clear that either a very large spaceship or multiple rockets will be needed to send the supplies to Mars which will be needed by a manned mission. A basic mission scenario for the first manned landing is provided here. This is from a report done with careful analysis of what would be needed for initial supplies and the following manned mission.

The following information is from the "MISSION TO MARS: HOW TO GET PEOPLE THERE AND BACK WITH NUCLEAR ENERGY" by the MIT Nuclear Engineering Department in 2003:

MANNED MISSIONS

Manned Missions: Launch Opportunity 1

In this initial manned mission launch opportunity, the cargo to be used by the first Mars exploration crew will be launched from Earth and sent, unmanned, to the Martian surface. It will be sent from Earth in two packages, CARGO1-A and CARGO1-B as described below. Once on

the surface, the infrastructure will deploy and activate under remote control. Mission controllers will monitor the health of these systems, and ensure that they are fully operational prior to committing the first crew on the next launch opportunity.

Suggested Launch Window: 2016 Launch Opportunity Objectives:

• To launch two additional Mars Transfer Systems (MTS2, MTS3).
• To send the first set of cargo (CARGO1-A & CARGO1-B) to the Martian surface.
• To provide an operational beacon to assist with the precision landing of future craft near the existing infrastructure.
• To ensure the delivered infrastructure is fully deployed and operational prior to the next launch opportunity.

Earth Launch Manifest:
• Magnum Heavy launch – MTS2 & Fuel
• Magnum Heavy launch – MTS3 & Fuel
• Magnum Heavy launch – CARGO1-A
• Magnum Heavy launch – CARGO1-B

Payload (CARGO1-A):

• Aero-capture shield for Mars orbital capture and entry.
• Surface landing system (retro-rockets, landing legs).
• Earth Entry / Mars Ascent Capsule.
• Surface Nuclear Fission Reactor Power System on mobile truck.
• 1 km long power cable.
• (ISRU) In-Situ Resource Utilization System.
• Hydrogen feedstock for ISRU.
• Surface communications system and landing assist beacon.
• Inflatable mars surface labs.
• Un-pressurized rover.

• Water storage tank.
• Science equipment.

Payload (CARGO1-B):

• Aero-capture shield for Mars orbital capture &
entry.
• Surface landing and repositioning system
(retro-rockets, landing legs and wheels).
• Surface Habitat

The opportunity can be subdivided into eight
stages which are:

- Launches to Low Earth Orbit
- MTS2 + MTS3 Activation
- CARGO1-A & MTS2 Docking
- CARGO1-B & MTS3 Docking
- Earth-Mars Transfers
- Aero Breaking & Landings
- CARGO1 Activation & Operations
- MTS2 & MTS3 Return

The resulting equipment ending up on Mars from
the first launches are depicted in the images
below:

Conceptual View of CARGO1-A Descent

Conceptual View of CARGO1-A Deployment

Conceptual View of CARGO1-B Descent

Manned Missions: Launch Opportunity 2

In the second launch opportunity, assuming that there is no problem with the initial set of infrastructure (CARGO1), two missions will occur. One will be the transfer of the first crew (CREW1) on a VASIMR driven fast transit (approx 90 days). The other will be the delivery of a second set of surface infrastructure (CARGO2-A and CARGO2-B) identical to the first set of cargo. The crew will land and utilize the existing infrastructure for life support and Mars

exploration.

Several months after crew arrival, CARGO2, which was on a slow efficient transfer, will land near the existing base and will serve as a backup habitat and set of infrastructure. Provided that this backup capability is not needed by CREW1, the second habitat and set of infrastructure will

be prepared for the next crew (CREW2). Following 600 days of surface activities, CREW1 will board their assent rocket and meet their

VASIMR craft in Mars orbit. The VASIMR craft will propel them back to Earth.

Suggested Launch Window: 2018 Launch Opportunity Objectives:

• To launch and assemble the first of two VASIMR Transfer Craft (VTC1).
• To launch the first crew (CREW1) and have them meet the VASIMR Transfer Craft in HEO.
• To send the first crew (CREW1) to Mars.
• To refuel MTS1 (sitting in LEO).
• To launch the fourth and final Mars Transfer System (MTS4).
• To send the second set of cargo (CARGO2-A and CARGO2-B) to the Martian surface.

Earth Launch Manifest:

• Magnum Heavy launch – VASIMR Transfer Craft 1 Section 1
• Magnum Heavy launch – VASIMR Transfer Craft 1 Section 2
• Delta 4 or Atlas 5 – CREW1 in Earth Ascent / Mars Descent Capsule
• Delta 4 or Atlas 5 – Xenon Refueler Craft
• Magnum Heavy launch – MTS4 and Fuel
• Magnum Heavy launch – CARGO2-A
• Magnum Heavy launch – CARGO2-B

Payload (VASIMR Transfer Craft Sections 1 and 2):

• 3 VASIMR engines
• 3 Space Nuclear Fission Reactors (4 MWe each)

- Hydrogen fuel for engines
- Crew Transfer Habitat
- Crew consumables (air, water, food)

Payload (Earth Ascent/Mars Descent Capsule):

- CREW1
- Ablative Mars Entry Shield.
- Surface landing system (retro-rockets, landing legs).
- Docking adapter.

Payload (Xenon Refueler):

- 25 metric tons of Xenon Fuel
- Fuel Tank Connector System
- Refueler RCS and maneuvering system.

Payload (CARGO2-A):
- Identical to CARGO1-A of 2016.

Payload (CARGO2-B):
- Identical to CARGO1-B of 2016.

The flow of the two missions taking place and what would be constructed on the surface of Mars is below:

Conceptual View of CREW1 Re-Positioning
Habitats

7.0 Necessary Life Support Systems

The following information from studies and International Space Station life support systems would also apply to a Mars transportation vehicle and to many of the systems inside a Mars colony.

7.1 Radiation Protection

On Earth we have a thick atmosphere to protect us from Gamma Rays in general and Solar Flares. In space none of this protection exists and we have to provide it.

On the International Space Station (ISS) In the current ISS, the materials for e.g. the hull are chosen primarily because they are light and strong. Aluminum is common. A few mm of Aluminum blocks most of the radiation you would encounter in low Earth orbit. In the ISS, 95% of the radiation is blocked.

This is enough for low Earth orbit: these orbits are inside the Van Allen Belts, so they are protected from the worst radiation. If we want to go beyond LEO for longer periods, more protection is needed. You could make the hull thicker, but this makes the launch more expensive

All About Mars Journeys and Settlement

7.2 Heat and Cold

The ISS has a lot of design elements used to maintain and control temperature. Without thermal controls, the temperature of the orbiting Space Station's Sun-facing side would soar to 250 degrees F (121 C), while thermometers on the dark side would plunge to minus 250 degrees F (-157 C). There might be a comfortable spot somewhere in the middle of the Station, but searching for it wouldn't be much fun!

Fortunately for the crew and all the Station's hardware, the ISS is designed and built with thermal balance in mind -- and it is equipped with a thermal control system that keeps the astronauts in their orbiting home cool and comfortable. The first design consideration for thermal control is insulation -- to keep heat in for warmth and to keep it out for cooling.

Here on Earth, environmental heat is transferred in the air primarily by conduction (collisions between individual air molecules) and convection (the circulation or bulk motion of air). "This is why you can insulate your house basically using the air trapped inside your insulation," said Andrew Hong, an engineer and thermal control specialist at NASA's Johnson Space Center. "Air is a poor conductor of heat, and the fibers of home insulation that hold the air still minimize convection."

"In space there is no air for conduction or convection," he added. Space is a radiation-dominated environment. Objects heat up by absorbing sunlight and they cool off by emitting infrared energy, a form of radiation which is invisible to the human eye.

As a result, insulation for the International Space Station doesn't look like the fluffy mat of pink fibers you often find in Earth homes. The Station's insulation is instead a highly-reflective blanket called Multi-Layer Insulation (or MLI) made of Mylar and Dacron.

The reflective silver mesh is aluminized Mylar. The copper-colored material is kapton, a heavier layer that protects the sheets of fragile Mylar, which are usually only 0.3 mil or 3/10000 of an inch thick.

"The Mylar is aluminized so that solar thermal radiation can't get through it," explains Hong. Here on Earth, we use blankets containing aluminized Mylar to wrap people who have been exposed to cold or trauma. Such blankets are especially popular among hunters and campers!

"Layers of Dacron fabric keep the Mylar sheets separated, which prevents heat from being conducted between layers," he continued. "This ensures radiation will be the most dominant heat transfer method through the blanket." Except for its windows, most of the ISS is covered with the radiation-stopping MLI.

There are also active heat tubes and thermal radiation fins to dump excess heat into space.

7.3 Fresh Water

All water used to be hauled into space by rocket then used up or wasted. The ISS now has a water production system in usage since 2010.

Drinkable water is one of the primary and most important assets for human survival. So when preparing for a journey, whether to sea or to space, planners must take this vital resource into consideration. Stowage space during such voyages always comes at a premium. It is no different for the International Space Station and the resupply vehicles that dock there.

A great example of a solution to minimize size and weight in life support is the recently launched Sabatier system. Originally developed by Nobel Prize-winning French chemist Paul Sabatier in the early 1900s, this process uses a catalyst that reacts with carbon dioxide and hydrogen - both byproducts of current life-support systems onboard the space station - to produce water and methane. This interaction closes the loop in the oxygen and water regeneration cycle. In other words, it provides a way to produce water without the need to transport it from Earth.

The fundamental technology for this particular system has been in development for the past twenty years. The overall schedule for hardware production, however, was under two years. This accelerated timeline was a significant challenge for the complex Sabatier, which contains a furnace, a multistage compressor, and a condenser/phase-separation system. The fact that recycling system feeds for Sabatier were already available on the station helped to simplify some of the design tasks by reducing the unknowns.

According to Jason Crusan, chief technologist for space operations at NASA Headquarters in Washington, the previous development and solid interfaces allowed NASA to try out a new way of acquiring services for the station with Sabatier. "Being able to demonstrate innovative new methods to acquire technical capabilities is one of the key cornerstones the space station can serve for future missions and approaches to those missions," Crusan explained.

Using developing technologies and productive systems enables the station to squeeze every drop from the resources that must launch from Earth. In addition to improving the efficiency of the station's resupply capabilities, Sabatier also frees up storage space. This helps to maximize the area available for science facilities and engineering equipment. The knowledge gained from such systems also advances the collective understanding of technologies to advance spaceflight and help solve similar problems on Earth.

The Sabatier system has long been a part of the space station plan, but the retirement of NASA's space shuttles elevated the need for new resources to provide water. For a decade, shuttles have provided water for the station as a byproduct of the fuel cells they use to generate electricity. Sabatier supplements the capability of resupply vehicles to provide water to the station, without becoming a sole source for this critical station resource.

Currently in operation on the station, Sabatier is the final piece of the regenerative environmental control and life-support system. This hardware was successfully activated in October 2010 and interacts directly with the Oxygen Generation System, which provides hydrogen, sharing a vent line.

Prior to Sabatier, the Oxygen Generation System vented excess carbon dioxide and hydrogen overboard. Rather than wasting these valuable chemicals, Sabatier enables their reuse to generate additional water for the station. With room and resources at a premium in space, this is a significant contribution to the space station's supply chain.

In addition there is now a degree of water recycling on the ISS. Nature's been recycling water on Earth for eons, and now NASA is set to do the same thing above Earth on the International Space Station. Space shuttle Endeavour carried in two refrigerator-sized racks packed with a distiller and an assortment of filters designed to process astronauts' urine and sweat into clean drinking water.

The station crew depends now on water carried up aboard a space shuttle or cargo rocket. But an operational water recycler is expected to cut that need by 65 percent by producing about 6,000 pounds of potable water each year. That's enough fresh water to allow the station to host six crew members instead of three.

A system that operates on the station also will provide a significant stepping stone to developing even more efficient processes that will support astronauts on the moon or on long-duration voyages into the solar system. Although Russia's space station Mir recycled cosmonaut's sweat, the NASA recycler is the first to be flown in space that intends to cleanse and reuse almost all the water a crew member produces.

The system can recycle about 93 percent of the water it receives, said Bob Bagdigian, the Environmental Control Life Support System project manager at NASA's Marshall Space Flight Center in Huntsville, Ala. The water recycler counts in large part on a distiller that Bagdigian compares to a keg tilted on its side. On Earth, distilling is a simple process of simply boiling water and cooling the steam back into pure water. But without gravity, the contaminants in water never separate from the steam no matter how much heat is used.

"In space, it becomes quite a challenge to distill any liquid in the absence of gravity," Bagdigian said.

So the keg-sized distiller is spun up to produce an artificial gravity field. The contaminants in the urine press against the sides of the drum while the steam gathers in the middle and is pumped to a filter. The filters are not much different from those used on Earth, which means they use charcoal-like materials to pull more unwanted elements from the water. Another process uses chemical compounds that bond with the remaining contaminants so filters can pick them out of the water, too.

"The water that we produce meets or exceeds most municipal water product standards," Bagdigian said. The system has been in different stages of development ever since NASA committed to building a space station in the 1980s. Along the way, individual parts of the system have been flown on space shuttle missions for tests.

The distiller mechanism flew in 2003 and worked just fine in orbit, Bagdigian said. Now the crew of the International Space Station will test the whole apparatus, but they won't drink any at first. Instead, they will take numerous samples and return them to Earth for detailed testing. After the testing is complete, controllers will clear the astronauts to use the fresh water in orbit.

NASA's water filter development has also helped produce filters that are now used in humanitarian efforts to make clean water in areas served only by contaminated sources. The effort to make a crew support system that reduces the need for fresh supplies from Earth includes an oxygen generator that is already installed in NASA's Destiny lab on the space station.

Housed in one rack instead of the two required for the water recycler, the oxygen producer splits the oxygen and hydrogen molecules in water and sends the oxygen into the space station as breathable air. The hydrogen is now dumped overboard. However, another process is under development that will combine the hydrogen with other chemicals that react with each other and produce more water.

While the water recycler in use will work fine for the International Space Station's needs, Bagdigian said work is already under way to make it more efficient so it can be used on long moon exploration missions. "We'll take this system and continue to push its performance and efficiency," Bagdigian said.

7.4 Breathable Atmosphere

Life support systems on the ISS must not only supply oxygen and remove carbon dioxide from the cabin's atmosphere, but also prevent gases like ammonia and acetone, which people emit in small quantities, from accumulating. Vaporous chemicals from science experiments are a potential hazard, too, if they combine in unforeseen ways with other elements in the air supply.

So, while air in space is undeniably rare, managing it is no small problem for ISS life support engineers. Most people can survive only a couple of minutes without oxygen, and low concentrations of oxygen can cause fatigue and blackouts. To ensure the safety of the crew, the ISS will have redundant supplies of that essential gas.

"The primary source of oxygen will be water electrolysis, followed by O2 in a pressurized storage tank," said Jay Perry, an aerospace engineer at NASA's Marshall Space Flight Center working on the Environmental Control and Life Support Systems (ECLSS) project. ECLSS engineers at Marshall, at the Johnson Space Center and elsewhere are developing, improving and testing primary life support systems for the ISS.

Most of the station's oxygen will come from a process called "electrolysis," which uses electricity from the ISS solar panels to split water into hydrogen gas and oxygen gas. Each molecule of water contains two hydrogen atoms and one oxygen atom. Running a current through water causes these atoms to separate and recombine as gaseous hydrogen (H2) and oxygen (O2).

The oxygen that people breathe on Earth also comes from the splitting of water, but it's not a mechanical process. Plants, algae, cyanobacteria and phytoplankton all split water molecules as part of photosynthesis -- the process that converts sunlight, carbon dioxide and water into sugars for food. The hydrogen is used for making sugars, and the oxygen is released into the atmosphere.

"Eventually, it would be great if we could use plants to (produce oxygen) for us," said Monsi Roman, chief microbiologist for the ECLSS project at MSFC. "The byproduct of plants doing this for us is food."

However, "the chemical-mechanical systems are much more compact, less labor intensive, and more reliable than a plant-based system," Perry noted. "A plant-based life support system design is presently at the basic research and demonstration stage of maturity and there are a myriad of challenges that must be overcome to make it viable."

Hydrogen that's left over from splitting water will be vented into space, at least at first. NASA engineers have left room in the ECLSS hardware racks for a machine that combines the hydrogen with excess carbon dioxide from the air in a chemical reaction that produces water and methane. The water would help replace the water used to make oxygen, and the methane would be vented to space.

"We're looking to close the loop completely, where everything will be (re)used," Roman said. Various uses for the methane are being considered, including expelling it to help provide the thrust necessary to maintain the Space Station's orbit. At present, "all of the venting that goes overboard is designed to be non-propulsive," Perry said.

The ISS also has large tanks of compressed oxygen mounted on the outside of the airlock module. These were the primary supply of oxygen for the U.S. segment of the ISS until the main life support systems arrived with Node 3 in 2005. After that, the tanks now serve as a backup oxygen supply.

All About Mars Journeys and Settlement

7.5 Pressure Containment

The main dangers to containing atmosphere on the ISS is for collisions with orbital debris. Even microscopic debris can be dangerous due to the speeds it is traveling of miles per second.

The key assets in Collision Avoidance are:

The telescopes and radars and satellites in the U.S. Department of Defense Space Surveillance Network that help in detecting, classifying and estimating orbital parameters of space debris computers in DoD's Joint Space Operations Center cranking through the large volumes of data obtained by the surveillance network and identifying dangerous stuff up there computers, qualified personnel and procedures at NASA Mission Control Center-Houston counterparts of the above in Russia (MCC-M, MSIC) When Debris Avoidance Maneuvers cannot be performed due to late detection of the threat, risk avoidance procedures force the ISS crew to go for the boats (err, Soyuz vehicles).

Mitigation comes into play after the collision. As discussed elsewhere (Are there any safety procedures in place on the ISS in case of puncture?), if there is time to isolate leaking compartments, the crew may do so. However, repairing the station is considered to be the job of follow-up expeditions. Broadly speaking, there are three other possible solutions to the problem that have not been implemented on the ISS:

Prevention of debris generation (by responsible design)

Debris collection

Active defense (with kinetic interceptors or laser ablation)

7.6 Waste Elimination

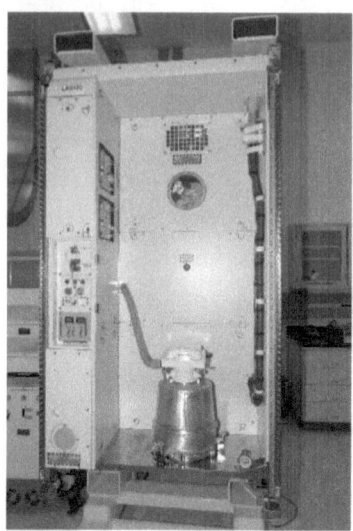

There are two interfaces used for the toilet. The hose, with the yellow funnel on the end, is used for urination. The commode with the circular aperture is used for defecating.

The astronaut lines themselves up properly and activates airflow that will direct the feces to go in the right direction once it leaves the body.

While liquid waste is indeed recycled into drinking water, solid waste is stored in a tank and that tank is periodically replaced. The full tank is put into a progress module which will burn up in the atmosphere.

All About Mars Journeys and Settlement

7.7 Power Systems

On the ISS there are lots of Solar Arrays to provide power all of the time.

Solar Array Wings

The electrical system of the International Space Station is a critical resource for the ISS because it allows the crew to live comfortably, to safely operate the station, and to perform scientific experiments. The ISS electrical system uses solar cells to directly convert sunlight to electricity. Large numbers of cells are assembled in arrays to produce high power levels. This method of harnessing solar power is called photovoltaics. The process of collecting sunlight, converting it to electricity, and managing and distributing this electricity builds up excess heat that can damage spacecraft equipment. This heat must be eliminated for reliable operation of the space station in orbit. The ISS power system uses radiators to dissipate the heat away from the spacecraft. The radiators are shaded from sunlight and aligned toward the cold void of deep space.

Each ISS solar array wing (often abbreviated "SAW") consists of two retractable "blankets" of solar cells with a mast between them. Each wing uses nearly 33,000 solar cells and when fully extended is 35 meters (115 ft.) in length and 12 meters (39 ft.) wide. When retracted, each wing folds into a solar array blanket box just 51 centimeters (20 in) high and 4.57 meters (15.0 ft.) in length. The ISS now has the full complement of eight solar array wings. Altogether, the arrays can generate 84 to 120 kilowatts.

The solar arrays normally track the Sun, with the "alpha gimbal" used as the primary rotation to follow the Sun as the space station moves around the Earth, and the "beta gimbal" used to adjust for the angle of the space station's orbit to the ecliptic. Several different tracking modes are used in operations, ranging from full Sun-tracking, to the drag-reduction mode ("Night glider" and "Sun slicer" modes), to a drag-maximization mode used to lower the altitude. See more details in the article at Night Glider mode.

Batteries

Since the station is often not in direct sunlight, it relies on rechargeable nickel-hydrogen batteries to provide continuous power during the "eclipse" part of the orbit (35 minutes of every 90 minute orbit). The batteries ensure that the station is never without power to sustain life-support systems and experiments. During the sunlit part of the orbit, the batteries are recharged. The nickel-hydrogen batteries have a design life of 6.5 years which means that they must be replaced multiple times during the expected 20-year life of the station. The batteries and the battery charge/discharge units are manufactured by Space Systems/Loral (SS/L), under contract to Boeing. N-H2 batteries on the P6 truss were replaced in 2009 and 2010 with more N-H2 batteries brought by Space Shuttle missions. There are batteries in Trusses P6, S6, P4, and S4.

Since 2017, nickel-hydrogen batteries are being replaced by lithium-ion batteries. On January 6, a multi-hour EVA began the process of converting some of the oldest batteries on the ISS to the new lithium-ion batteries There are a number of differences between the two battery technologies, and one difference is that the lithium-ion batteries can handle twice the charge, so only half as many lithium-ion batteries are needed during replacement. Also, the lithium-ion batteries are smaller than the older nickel-hydrogen batteries. Although they are not quite as long lasting as nickel-hydrogen, they can last long enough to extend the life of ISS.

ISS Electrical Power Distribution

The power management and distribution subsystem operates at a primary bus voltage set to Vmp, the peak power point of the solar arrays. As of 30 December 2005, Vmp was 160 volts DC (direct current). It can change over time as the arrays degrade from ionizing radiation. Microprocessor-controlled switches control the distribution of primary power throughout the station.

The battery charge/discharge units (BCDUs) regulate the amount of charge put into the battery. Each BCDU can regulate discharge current from two battery ORUs (Orbital Replacement Unit, a series-connected pack of 38 Ni-H2 cells), and can provide up to 6.6 kW to the Space Station. During insolation, the BCDU provides charge current to the batteries and controls the amount of battery overcharge. Each day, the BCDU and batteries undergo sixteen charge/discharge cycles. The Space Station has 24 BCDUs, each weighing 100 kg.

Data Architecture/Communications

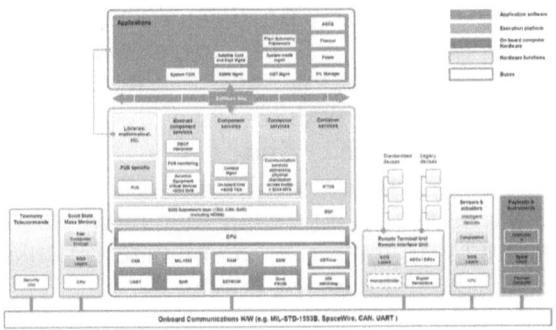

The ISS data architecture and communications system is very complex. I included an architecture diagram above and detailed overview below so that you can see just how much is involved.

On a larger habitat in space imagine that the architecture is that much more complex according to its size and the number of people on it. Fortunately, computing architecture is one area where continuous advances should keep up with the computing needs of an advanced space habitat facility.

Spacecraft Management Unit

On the ISS The On-board Computer (also referred to as Spacecraft Management Unit - SMU or Command & Data Handling Management unit - CDMU) is the central core of the Spacecraft Avionics. The Central Processing Unit (CPU) hosts the Execution Platform SW (composed of RTOS, BSP, SOIS layers, PUS, ...) and the Application SW. Volatile and Non-volatile Memories, Safe Guard Memories, On Board Timer, Interface controllers and Reconfiguration modules are the other main blocks of a OBC. The figure above shows a functional architecture of the On-Board Data System where all the major functional blocks are indicated with their intercommunication links and their typical redundancy scheme.

Remote Terminal Unit

Remote Terminal Unit (also called Remote Interface Unit-RIU) is a unit that is usually present on medium-large size spacecraft. The RTU offloads the On Board Computer from analogue and discrete digital data acquisition and actuators control tasks.

Platform Solid State Mass Memory

For Earth Observation missions the mass memory for the P/L data may belong to the satellite platform and sometimes, depending on the capacity required, might be included inside the OBC as a single module.

TM/TC

The tele commands, once validated, are multiplexed to the intended addresses. There are two categories of commands: the high priority and the normal commands. The high priority commands (HPC) are sent to the Command Pulse Distribution Unit (CPDU) for immediate execution. The CPDU is either internal to the TC decoder or external and it's implemented in hardware, i.e. no software is involved in the execution of HPCs. The normal commands are sent off to the OBC CPU to be either processed or relayed on the system bus. The Telemetry encoder collects the Telemetry packets from different sources (processing, data storage, essential telemetry, payload), assembles the Telemetry transfer frames and sends them to the TM/TC transceiver to be downloaded to the ground.

Busses

The most common command and control bus used on a spacecraft platform is the MIL-STD-1553B covered by the ECSS-E-ST-50-13C. An alternative to the MIL-STD-1553B is the CAN that ESA and the European Space community is standardizing for space applications. UART serial channels are also used especially to control AOCS sensors. The Spacewire technology is now being increasingly used for data transfers < 160 Mbit/s and it can combine the command and control function with massive data transfer.

All About Mars Journeys and Settlement

7.8 Communication protocols

The space community is asking for a real improvement in the specification and use of communications protocols. Typically, previous developments have harmonized physical interfaces and low level data link protocols but above this level proprietary solutions have been utilized. This has without any doubt increased development and integration costs and limited the possibility of element reuse without expensive modification. In comparison, the commercial market on the ground has systematically pursued the use of multilayer protocol stacks resulting in simple integration and multi-vendor compatibility. This commercial trend is now being adopted for the flight avionics by the development and standardization of protocols above the basic link layer.

7.9 Food Production

Growing food in space only makes sense. It takes too much energy to ship all your food into space. Also, plants produce oxygen to help with the atmosphere in a habitat. One astronaut on the International Space Station requires approximately 1.8 kilograms of food and packaging per day. For a long-term mission, such as a four-man crew, three year Martian mission, this number can grow to as much as 24,000 pounds.

Due to the cost of resupply and the impracticality of resupplying interplanetary missions the prospect of growing food inflight is incredibly appealing. The existence of a space farm would aid the creation of a sustainable environment, as plants can be used to recycle wastewater, generate oxygen, continuously purify the air and recycle faeces on the space station or spaceship. Just 10m² of crops produces 25% of the daily requirements of 1 person, or about 180-210grams of oxygen. This essentially allows the space farm to turn the spaceship into an artificial ecosystem with a hydrological cycle and nutrient recycling.

In addition to maintaining a shelf-life and reducing total mass, the ability to grow food in space would help reduce the vitamin gap in astronaut's diets and provide fresh food with improved taste and texture. Currently, much of the food supplied to astronauts is heat treated or freeze dried. Both of these methods, for the most part, retain the properties of the food pre-treatment. However, vitamin degradation during storage can occur. A 2009 study noted significant decreases in vitamins A, C and K as well as folic acid and thiamin can occur in as little as one year of storage. A mission to Mars could require food storage for as long as five years, thus a new source of these vitamins would be required.

Supply of foodstuffs to others is likely to be a major part of early off-Earth settlements. Food production is a non-trivial task and is likely to be one of the most labor-intensive, and vital, tasks of early colonists. Among others, NASA is researching how to accomplish space farming.

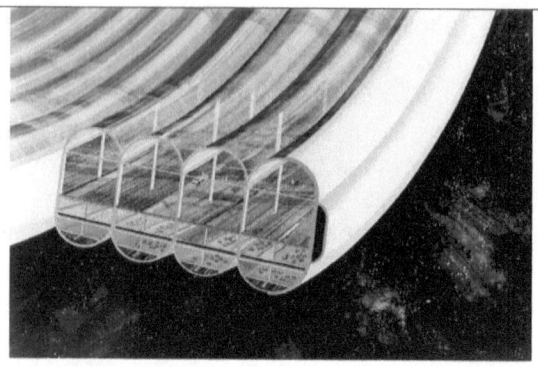

More space food production studies:

ASA plant physiologist Ray Wheeler, Ph.D., and fictional astronaut Mark Watney from the movie "The Martian" have something in common — they are both botanists. But that's where the similarities end. While Watney is a movie character who gets stranded on Mars, Wheeler is the lead for Advanced Life Support Research activities in the Exploration Research and Technology Program at Kennedy Space Center, working on real plant research.

"The Martian movie and book conveyed a lot of issues regarding growing food and surviving on a planet far from the Earth," Wheeler said. "It's brought plants back into the equation."

As NASA prepares the Space Launch System rocket and Orion spacecraft for Exploration Mission-1, it's also turning its attention to exploring the possibilities of food crops grown in controlled environments for long-duration missions to deep-space destinations such as Mars.

Wheeler and his colleagues, including plant scientists, have been studying ways to grow safe, fresh food crops efficiently off the Earth. Most recently, astronauts on the International Space Station harvested and ate a variety of red romaine lettuce that they activated and grew in a plant growth system called Veggie.

Wheeler, who has worked at Kennedy since 1988, was among the plant scientists and collaborators who helped get the Veggie unit tested and certified for use on the space station. The plant chamber, developed by Orbitec through a NASA Small Business Innovative Research Program, passed safety reviews and met low power usage and low mass requirements for use on the space station.

Aside from the chamber, the essentials needed for growing food crops, whether on the Earth or another planet, such as Mars, are water, light and soil, along with some kind of nutrients to help them grow.

Potato Crop Studies

A variety of red potatoes called Norland were grown in the Biomass Production Chamber inside Hangar L at Cape Canaveral Air Force Station in Florida during a research study in 1992.

What kind of crops could be grown in space or on another planet? Potatoes, sweet potatoes, wheat and soybeans would all be good according to Wheeler because they provide a lot of carbohydrates, and soybeans are a good source of protein.

Also, potatoes are tubers, which means they store their edible biomass in underground structures. Wheeler said potatoes could produce twice the amount of food as some seed crops when given equivalent light. After salad crops that are now being studied, they are the next category of minimally processed food crops and could be consumed raw.

"You could begin to grow potatoes, wheat and soybeans, things like that, and along with the salad crops, you could provide more of a complete diet," Wheeler said.

Wheeler has spent a lot of time studying different ways to grow potatoes. Most of his studies took place during the late 1980s through the early 2000s inside Hangar L at Cape Canaveral Air Force Station in Florida. The lab was relocated to the Space Life Sciences Laboratory in 2003. A major portion of the labs were then relocated to the Space Station Processing Facility in 2014 to become part of the Exploration Research and Technology Programs Directorate at Kennedy.

Many of the early potato crop studies were done at the University of Wisconsin, where Wheeler worked prior to coming to Kennedy. Plant scientists at Kennedy used these fundamental findings as a starting point for their studies, and in particular, a variety called Norland red potatoes, using a large plant chamber called the Biomass Plant Production Chamber.

The Biomass Production Chamber originally was a hypobaric test chamber used during the Mercury Project. Including its pedestal, the chamber is 28 feet tall. It was later modified to grow plants in the mid-1980s. Air circulation ducts and fans, high pressure sodium lamps, cooling and heating systems, and hydroponic trays and solution tanks were added. The chamber provided a tightly closed atmosphere for plant growth, which simulated what might be encountered in space.

"Providing food is a complex issue," Wheeler said. "We have to think about nutritional issues, what's acceptable and what tastes good. If nobody wants to eat it, that won't work."

7.10 Atmosphere

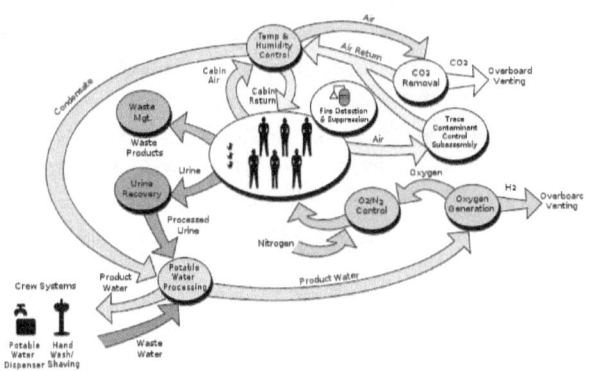

On the ISS the air has to be circulated continuously and everywhere in the station to make sure it is all breathable. This will become an even bigger issue on larger space habitats.

The International Space Station Environmental Control and Life Support System (ECLSS) is a life support system that provides or controls atmospheric pressure, fire detection and suppression, oxygen levels, waste management and water supply. The highest priority for the ECLSS is the ISS atmosphere, but the system also collects, processes, and stores waste and water produced and used by the crew—a process that recycles fluid from the sink, shower, toilet, and condensation from the air.

The Elektron system aboard Zvezda and a similar system in Destiny generate oxygen aboard the station. The crew has a backup option in the form of bottled oxygen and Solid Fuel Oxygen Generation (SFOG) canisters.

Carbon dioxide is removed from the air by the Russian Vozdukh system in Zvezda, one Carbon Dioxide Removal Assembly (CDRA) located in the U.S. Lab module, and one CDRA in the U.S. Node 3 module. Other by-products of human metabolism, such as methane from the intestines and ammonia from sweat, are removed by activated charcoal filters or by the Trace Contaminant Control System (TCCS).

8.0 Psychological Issues

Mission to Mars

What psychosocial challenges would astronauts face on an epic journey to the red planet?

Even at their closest, when Earth and Mars are approaching one another in their oblong orbits, there are 35 million miles between our blue orb and the red planet. But that distance hasn't stopped NASA and other space agencies from setting their sights on a human mission to Mars, which would require astronauts to live in space for at least two and a half years. NASA has been working toward the goal of delivering astronauts to Mars by the early 2030s, says Thomas Williams, PhD, a psychologist and chief scientist of human factors and behavioral performance at NASA's Johnson Space Center in Houston, Texas. In December, President

Donald J. Trump signed a space policy directive instructing the space agency to return astronauts to the moon. It's been over 50 years since astronauts last traveled beyond Earth's orbit, and a fresh moon mission would serve as a pock-marked steppingstone toward a subsequent human mission to Mars.

Getting there requires not only rocket science, but also human science—which can be even more daunting than perfecting propulsion systems and landing gear.

NASA doesn't take these challenges lightly, Williams says. Scientists are carefully assessing the physiological, psychological and social factors associated with a journey to Mars, with NASA conducting research independently as well as in partnership with experts outside of the agency. "We are concerned about the risks behavioral health and performance would pose to a Mars mission within our current understanding, but we have cutting-edge researchers to help us answer these questions," he says. "None of this we consider insurmountable."

Far from home

Scientists have been studying the effects of living in space for decades. Since 1971, astronauts (and their Russian counterparts, cosmonauts) have spent weeks or months inhabiting a series of space stations orbiting Earth. The current iteration, the International

Space Station (ISS), welcomed its first crew in 2000 and has been occupied ever since.

Research from the space station has provided useful information about how astronauts respond to the challenges of space, such as microgravity, confinement and isolation, says Nick Kanas, MD, emeritus professor of psychiatry at the University of California, San Francisco, who has long studied space psychology. But interplanetary travel is another story, he says. "Mars is a long way away, and the extreme distance has psychological ramifications."

At the Hawai'i Space Exploration Analog and Simulation habitat astronauts spend about six months at a time on the ISS. So far, the longest consecutive period spent in space is 437 days, a record set by cosmonaut Valeri Polyakov aboard the Russian station Mir. A multiyear journey is uncharted territory—and a long time to spend in tight quarters with just three or four other people. "It will be hard to have the kind of social novelty we crave," Kanas says. And because a Mars expedition would probably be a collaborative effort among countries, those astronauts will have to overcome cultural differences to live and work together.

What's more, communication between Earth and a Mars-bound ship will be delayed up to 20 minutes each way. If an astronaut asked a question, 40 minutes could pass before he or she received a reply. "We know the ability to talk in real time with family and with the people on mission control is very important to astronauts," Kanas says. "When you take that away, it creates a real conflict."

The difficulty of speaking with family members on the ground could contribute to loneliness and psychological problems such as anxiety or depression. Astronauts will also have to be much more autonomous and prepared to handle emergencies on their own, since they won't be able to rely on real-time advice from mission control. "The communication delay will lead to crews having to take care of their own problems, including medical or psychological emergencies," Kanas says.

Another concern is how astronauts might react to the experience of being so far from Earth. Many ISS astronauts report that gazing at and photographing the Earth from above is a favorite pastime that can reduce stress and even inspire spiritual or transcendental experiences, as Kanas described in a review of the psychosocial issues related to long-distance space travel (Acta Astronautica, Vol. 103, No. 1, 2014). But that perk won't be available from 35 million miles away. "Nobody knows the effect of seeing the Earth as a dot in the heavens," he says. "Maybe it won't have any effect—but maybe it will."

Floating questions

The physical challenges of living in space can also have implications for psychological well-being. One top concern is how space radiation will affect the body. Beyond the protective bubble of Earth's atmosphere, space radiation poses a significant threat to human DNA, cells and tissues. "It can impact the central nervous system and can alter the structure and function of the brain," says Williams.

In animal studies, NASA scientists are exploring how chronic radiation exposure might affect brain function. Recently, Charles Limoli, PhD, at the University of California, Irvine, and colleagues exposed mice to charged particles that simulated cosmic radiation. They found structural changes in the mice's brains such as reduced complexity of dendrites, the extensions that branch from neurons. What's more, the mice also showed behavioral changes, including memory deficits, increased anxiety and deficits in executive function (Science Reports, Vol. 6, No. 1, 2016).

Researchers are also studying factors that might compound or minimize those effects, Williams says. For example, scientists are exploring the concept of cognitive reserve, which posits that education and experience can help protect the brain against physical damage (such as the pathological changes associated with Alzheimer's disease). NASA scientists are investigating whether cognitive reserve can also protect against radiation in space.

The effect of altered gravity is another active area for space scientists. Floating weightless might look like fun, but it can lead to physical problems including motion sickness, muscle wasting and changes in visual perception. Those changes could have downstream effects on psychological well-being, Williams notes. Getting regular exercise, for instance, is a lot trickier when your feet don't touch the ground—and physical activity is known to promote positive mental health.

Weightlessness can also contribute to psychological problems in surprising ways. For example, kidney stones are more common in altered gravity environments, and stones can raise the risk of urinary tract infections, Williams says. In some cases, undiagnosed UTIs can trigger confusion or delirium, which could be mistaken for a psychiatric disorder. Researchers' work to connect the dots among these possible risks is essential for preparing astronauts for life in space. "It's important to be alert to what medical conditions could occur as a result of these altered environments," Williams says. "If someone has a medical condition, we don't want to treat it as a psychological problem."

Mental Health on Mars

Prolonged weightlessness is hard to study on Earth, where it's impossible to cancel out the effects of our planet's gravity. But many other features of an extended space mission can be recreated in so-called space analog studies conducted in confined and isolated environments.

U.S. Air Force Reserve Officer Casey Stedman served as commander for the second mission of the Hawai'i Space Exploration Analog and Simulation project. The largest such study to date was the Mars500 project, led by the Institute of Biomedical Problems of the Russian Academy of Sciences in 2010–11. For 520 days, six healthy male participants from several countries lived inside an enclosed module in Russia designed to mimic the feel and function of a Mars shuttle. Crew members had military and engineering backgrounds, similar to the traditional backgrounds of astronauts and cosmonauts. During the simulation, the crew members performed routine maintenance and scientific experiments, were isolated from Earth's light-dark cycles and experienced communication delays just as they would on a flight to Mars.

That experiment raised some concerns, says David Dinges, PhD, a psychologist at the University of Pennsylvania who researches chronobiology and has studied astronauts on the ISS and in space analog environments. He and his colleagues recorded psychological and behavioral changes among the Mars500 participants. One crew member experienced mild to moderate symptoms of depression during most of his time in confinement, they found. Two others experienced abnormal sleep-wake cycles, while another reported insomnia and physical exhaustion (PLOS One, Vol. 9, No. 3, 2014).

In the same study, he and his colleagues also found that the two crew members who had the highest rates of stress and exhaustion were involved in 85 percent of the perceived conflicts with other crew members and mission control. A single stressed-out astronaut, in other words, might cause problems that affect the entire mission.

In a different study, Dinges and colleagues looked more closely at the sleep and activity habits of the Mars500 crew. They found that as the months stretched on, crew members became increasingly sedentary when awake. They also spent more time sleeping and resting, which the researchers characterized as a kind of behavioral "torpor" to conserve energy. Four of the six crew members experienced sleep problems during their 520 days in pseudospace, including disrupted sleep-wake schedules, reduced sleep quality, a shift to more daytime sleep and performance deficits related to chronic sleep loss (PNAS, Vol. 110, No. 7, 2013).

Some of those problems can be addressed by optimizing lighting to more accurately mimic the 24-hour cycle and UV spectrum of sunlight on Earth, Dinges says. "We are a circadian species, and if you don't have the proper lighting to maintain that chronobiology, it can create significant problems for crew members," he explains. Other elements, such as stress or operational factors related to crew work schedules, might also be to blame, he says. But more research is needed to fully understand those factors.

Coping skills

Dinges is among a large team of scientists working to understand and prevent psychosocial problems that might arise in space. In a new NASA-supported project, he and colleagues at institutions across the country are looking for biological indicators that offer clues about a person's emotional, social and cognitive resilience. "We know there are substantial individual differences in how people cope with different kinds of stressors, but we don't understand why that is—and more importantly, how to identify ahead of time how people might cope, or how to help them do it," he says.

NASA already employs a comprehensive physical and psychological screening process to identify astronauts likely to thrive under the stressful conditions of spaceflight. Biomarkers of resilience could add a new dimension to that evaluation. However, selecting biologically superior astronauts isn't necessarily the goal, says Dinges. Instead, he envisions such biomarkers being used in research to identify and test medications or behavioral strategies that could boost resilience.

"Biomarkers could be used to determine how to maximize those countermeasures," he says. "Behavioral issues are serious, and the challenge isn't just to figure out who can optimally cope, but also how to provide help for those who need it."

It's too soon to say whether scientists will succeed in finding a blood test to measure resilience. But with or without such a test, a Mars mission requires a strategy to help astronauts manage stress and maintain emotional well-being.

Raphael Rose, PhD, a psychologist at the University of California, Los Angeles, is among the scientists contributing to that effort. He has studied a stress management program among participants in the Hawai'i Space Exploration Analog and Simulation (HI-SEAS) project, a study led by the University of Hawai'i at Mānoa. During the fifth installment of the project in 2017, six men and women spent eight months living and working in an isolated compound on the rocky, otherworldly landscape of Mauna Loa.

During that project, participants used Rose's program, the Stress Management and Resilience Training for Optimal Performance (SMART-OP). The program involves a variety of self-guided modules, such as video demonstrations of conflict-resolution strategies and a biofeedback game that helps the user practice regulating his or her breathing and heart rate. Rose says NASA is reviewing the findings, which have not yet been released to the public. But he's optimistic the program shows promise. "The participants found the program really useful and helpful, which is a good sign that it's something people would actually use, and something that could be integrated into future trainings or missions."

"These crews contain rather remarkable people who are already quite resilient to start with," Rose says. "But we can lower the risk of any behavioral health concerns by addressing things in advance through training and providing people with an avenue to address any issues that come up."

Studies such as Mars500 and HI-SEAS are important, but can't answer all of the questions about life in space. Sample sizes are tiny. And participants know they're not actually hurtling through the void of space. Of course, there's that pesky thing called gravity.

Yet while a true trip to Mars might have unique stressors, it is also likely to be filled with excitement and wonder. For people who have long dreamed of exploring the next frontier, those benefits are likely to outweigh the potential risks. "The reality is it would be pretty exciting for astronauts to actually be the first to walk on Mars," Williams says.

9.0 Radiation Protection

One of the biggest unsolved issues about a journey to Mars and living on Mars is radiation. The effects will be worst in inter solar space and living on the surface of Mars than any situation man has yet experienced.

Shielding the humans

The biggest threat to human life on a Mars-bound mission may be the one we can't see – radiation. "It's certainly one of the biggest technical challenges," says Parker. "Although the space-station astronauts are exposed to more radiation than we are on the ground, they're substantially protected by the Van Allen belts." These giant rings of charged particles, which originate from the interaction of the magnetic field and the solar wind, shield Earth from high-energy particles.

Going to the Moon or to Mars, however, means leaving the security of the magnetosphere behind. In deep space, radiation from sources such as solar activity and galactic cosmic rays, includes high-energy protons, photons and electrons. These particles can damage shields and penetrate many materials, including human skin. The harmful effects of space radiation accumulate over time, which means Mars-bound explorers on a three-year mission will be exposed to a radiation dose more than 100 times greater than the average dose received by Apollo astronauts, who were outside the magnetosphere for only a few days.

As Mars' thin atmosphere doesn't offer much protection, previous and ongoing missions to Mars have been vital in the design of materials that will keep Martian-explorers safe not only during the journey, but while they're there. In a study published in 2014 Scientists used data collected by the Radiation Assessment Detector (RAD) on NASA's Curiosity rover – the first radiation analysis done on another planet – to estimate that 500 days on Mars would expose people to more than 120 mSv of radiation from the Sun and cosmic rays. That's more than 100 times the average annual radiation exposure on Earth. When they include radiation measurements collected during the travel to and from Mars, the researchers estimate total exposure could be as much as 1 Sv. Previous studies estimate that that much radiation would raise a 40-year-old man's risk of dying from cancer to at least 4%, and a 40-year-old woman's risk to 5%.

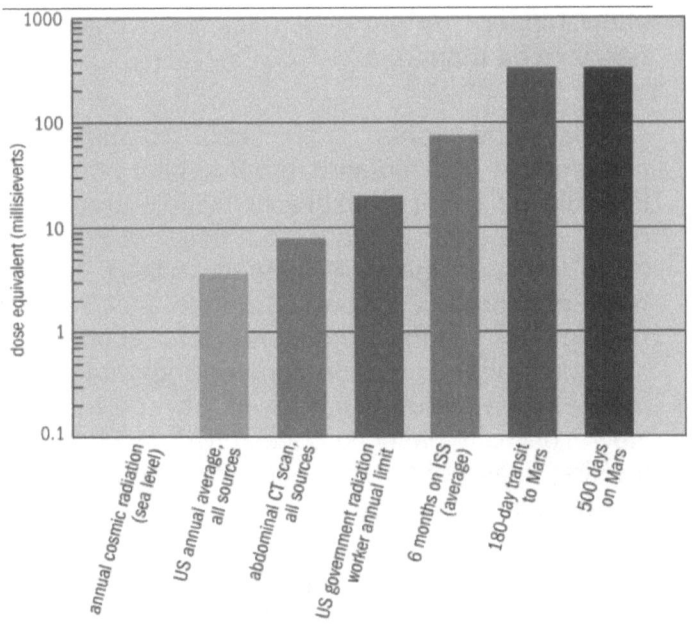

Comparing doses

Before RAD, physicists had to rely on computer models to predict the radiation environment. "But as a good physicist, you need to do the experiments," says Don Hassler of Southwest Research Institute in Boulder, Colorado. "You can't believe your models until you've validated them." As it turned out, existing models didn't match the details of RAD's data and had to be improved. Hassler, who led the 2014 study, says RAD's data revealed unexpected variations in the radiation on the Martian surface, not only seasonally but also day to day. "The radiation environment is not a showstopper," he says. "It's like the weather. It needs to be managed."

Real-world data will also inform the construction of a radiation-proof habitat on Mars. It would be cost-prohibitive to actually transport building materials to Mars, so a better tack would be to transport the tools instead. Indeed, NASA has sponsored a $2.5m challenge that invites people to submit designs for 3D-printed structures that could be forged from materials native to the planet, such as ice or regolith. The New York City-based architecture firm that won the first phase of the contest submitted a design of a house made of ice.

Our explorers will need to have a safe spaceship and quarters on Mars to let them only experience safe radiation doses. This could include some type of structure where water is shielding the habitation module on the trip there. On Mars building shelters underground or covering the buildings with soil will have the same effectiveness of protecting against the radiation dosages.

10.0 Commercial Aspects

10.1 Economics

As with early colonies in the New World, economics would be a crucial aspect to a colony's success. The reduced gravity well of Mars and its position in the Solar System may facilitate Mars–Earth trade and may provide an economic rationale for continued settlement of the planet. Given its size and resources, this might eventually be a place to grow food and produce equipment to mine the asteroid belt.

Some early Mars colonies might specialize in developing local resources for Martian consumption, such as water and/or ice. Local resources can also be used in infrastructure construction. One source of Martian ore currently known to be available is metallic iron in the form of nickel–iron meteorites. Iron in this form is more easily extracted than from the iron oxides that cover the planet.

Another main inter-Martian trade good during early colonization could be manure. Assuming that life doesn't exist on Mars, the soil is going to be very poor for growing plants, so manure and other fertilizers will be valued highly in any Martian civilization until the planet changes enough chemically to support growing vegetation on its own.

Solar power is a candidate for power for a Martian colony. Solar insolation (the amount of solar radiation that reaches Mars) is about 42% of that on Earth, since Mars is about 52% farther from the Sun and insolation falls off as the square of distance. But the thin atmosphere would allow almost all of that energy to reach the surface as compared to Earth, where the atmosphere absorbs roughly a quarter of the solar radiation. Sunlight on the surface of Mars would be much like a moderately cloudy day on Earth.

10.2 Economic drivers

Space colonization on Mars can roughly be said to be possible when the necessary methods of space colonization become cheap enough (such as space access by cheaper launch systems) to meet the cumulative funds that have been gathered for the purpose.

Although there are no immediate prospects for the large amounts of money required for any space colonization to be available given traditional launch costs, there is some prospect of a radical reduction to launch costs in the 2020s, which would consequently lessen the cost of any efforts in that direction. With a published price of US$62 million per launch of up to 22,800 kg (50,300 lb) payload to low Earth orbit or 4,020 kg (8,860 lb) to Mars, SpaceX Falcon 9 rockets are already the "cheapest in the industry". SpaceX's reusable plans include Falcon Heavy and future methane-based launch vehicles including the Starship. If SpaceX is successful in developing the reusable technology, it would be expected to "have a major impact on the cost of access to space", and change the increasingly competitive market in space launch services.

Alternative funding approaches might include the creation of inducement prizes. For example, the 2004 President's Commission on Implementation of United States Space Exploration Policy suggested that an inducement prize contest should be established, perhaps by government, for the achievement of space colonization. One example provided was offering a prize to the first organization to place humans on the Moon and sustain them for a fixed period before they return to Earth.

11.0 Earth-Mars Cycler Ideas

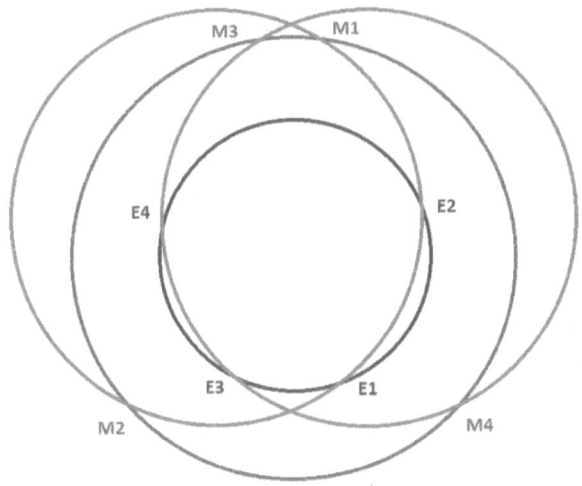

A diagram of the Aldrin cycler with two different
transfer orbits

A cycler is a trajectory that encounters two or more bodies regularly. Once the orbit is established, no propulsion is required to shuttle between the two, although some minor corrections may be necessary due to small perturbations in the orbit. The use of cyclers was considered in 1969 by Walter M. Hollister, who examined the case of an Earth–Venus cycler. Hollister did not have any particular mission in mind, but posited their use for both regular communication between two planets, and for multi-planet flyby missions.

A Martian year is 1.8808 Earth years, so Mars makes eight orbits of the Sun in about the same time as Earth makes 15. Cycler trajectories between Earth and Mars occur in whole-number multiples of the synodic period between the two planets, which is about 2.135 Earth years. In 1985, Buzz Aldrin presented an extension of his earlier Lunar cycler work which identified a Mars cycler corresponding to a single synodic period. The Aldrin cycler (as it is now known) makes a single eccentric loop around the Sun. It travels from Earth to Mars in 146 days (4.8 months), spends the next 16 months beyond the orbit of Mars, and takes another 146 days going from the orbit of Mars back to the first crossing of Earth's orbit.

The existence of the now-eponymous Aldrin Cycler was calculated and confirmed by scientists at Jet Propulsion Laboratory later that year, along with the VISIT-1 and VISIT-2 cyclers proposed by John Niehoff in 1985. For each Earth–Mars cycler that is not a multiple of 7 synodic periods, an outbound cycler intersects Mars on the way out from Earth while an inbound cycler intersects Mars on the way in to Earth. The only difference in these trajectories is the date in the synodic period in which the vehicle is launched from Earth. Earth–Mars cyclers with a multiple of 7 synodic periods return to Earth at nearly the same point in its orbit and may encounter Earth and/or Mars multiple times during each cycle. VISIT 1 encounters Earth 3 times and Mars 4 times in 15 years. VISIT 2 encounters Earth 5 times and Mars 2 times in 15 years.

12.0 Does Mars have Life on it?

The idea that life exists on Mars has been debated at least since the nineteenth century. One compound which exists on Mars in abundance is Water.

Living Water

Liquid water is a necessary but not sufficient condition for life as humans know it, as habitability is a function of a multitude of environmental parameters. Liquid water cannot exist on the surface of Mars except at the lowest elevations for minutes or hours. Liquid water does not appear at the surface itself, but it could form in minuscule amounts around dust particles in snow heated by the Sun. Also, the ancient equatorial ice sheets beneath the ground may slowly sublimate or melt, accessible from the surface via caves.

Mars - Utopia Planitia

Scalloped terrain led to the discovery of a large amount of underground ice enough water to fill Lake Superior (November 22, 2016)

Martian terrain

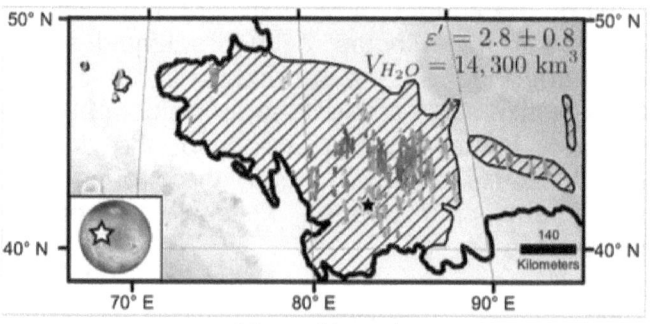

$$\varepsilon' = 2.8 \pm 0.8$$
$$V_{H_2O} = 14,300 \text{ km}^3$$

Map of terrain

Water on Mars exists almost exclusively as water ice, located in the Martian polar ice caps and under the shallow Martian surface even at more temperate latitudes. A small amount of water vapor is present in the atmosphere. There are no bodies of liquid water on the Martian surface because its atmospheric pressure at the surface averages 600 pascals (0.087 psi)—about 0.6% of Earth's mean sea level pressure—and because the temperature is far too low, (210 K (−63 °C)) leading to immediate freezing. Despite this, about 3.8 billion years ago, there was a denser atmosphere, higher temperature, and vast amounts of liquid water flowed on the surface, including large oceans.

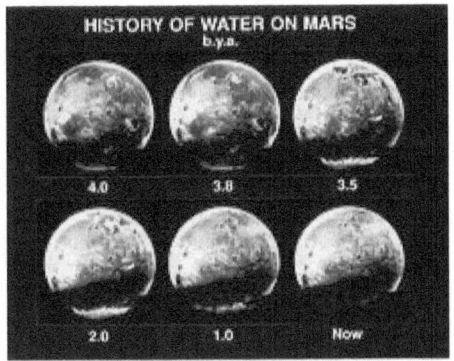

A series of artist's conceptions of past water coverage on Mars

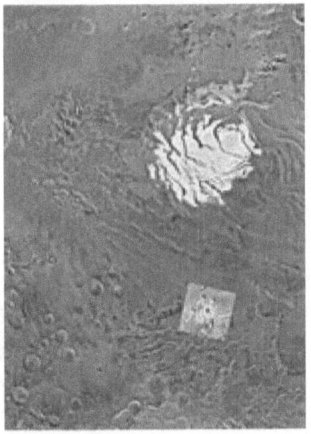

Mars South Pole

Site of Subglacial Water (July 25, 2018)

It has been estimated that the primordial oceans on Mars would have covered between 36% and 75% of the planet. On November 22, 2016, NASA reported finding a large amount of underground ice in the Utopia Planitia region of Mars. The volume of water detected has been estimated to be equivalent to the volume of water in Lake Superior. Analysis of Martian sandstones, using data obtained from orbital spectrometry, suggests that the waters that previously existed on the surface of Mars would have had too high a salinity to support most Earth-like life. Tosca et al. found that the Martian water in the locations they studied all had water activity, $a_w \leq 0.78$ to 0.86—a level fatal to most Terrestrial life. Haloarchaea, however, are able to live in hypersaline solutions, up to the saturation point.

In June 2000, possible evidence for current liquid water flowing at the surface of Mars was discovered in the form of flood-like gullies. Additional similar images were published in 2006, taken by the Mars Global Surveyor that suggested that water occasionally flows on the surface of Mars. The images showed changes in steep crater walls and sediment deposits, providing the strongest evidence yet that water coursed through them as recently as several million years ago.

There is disagreement in the scientific community as to whether or not the recent gully streaks were formed by liquid water. Some suggest the flows were merely dry sand flows. Others suggest it may be liquid brine near the surface, but the exact source of the water and the mechanism behind its motion are not understood.

In July 2018, scientists reported the discovery of a subglacial lake on Mars, 1.5 km (0.93 mi) below the southern polar ice cap, and extending sideways about 20 km (12 mi), the first known stable body of water on the planet. The lake was discovered using the MARSIS radar on board the Mars Express orbiter, and the profiles were collected between May 2012 and December 2015. The lake is centered at 193°E, 81°S, a flat area that does not exhibit any peculiar topographic characteristics but is surrounded by higher ground, except on its eastern side, where there is a depression.

Silica

This silica-rich patch discovered by Spirit rover In May 2007, the Spirit rover disturbed a patch of ground with its inoperative wheel, uncovering an area 90% rich in silica. The feature is reminiscent of the effect of hot spring water or steam coming into contact with volcanic rocks.

Scientists consider this as evidence of a past environment that may have been favorable for microbial life and theorize that one possible origin for the silica may have been produced by the interaction of soil with acid vapors produced by volcanic activity in the presence of water.

Based on Earth analogs, hydrothermal systems on Mars would be highly attractive for their potential for preserving organic and inorganic bio signatures. For this reason, hydrothermal deposits are regarded as important targets in the exploration for fossil evidence of ancient Martian life.

Possible Bio signatures

In May 2017, evidence of the earliest known life on land on Earth may have been found in 3.48-billion-year-old geyserite and other related mineral deposits (often found around hot springs and geysers) uncovered in the Pilbara Craton of Western Australia. These findings may be helpful in deciding where best to search for early signs of life on the planet Mars.

Methane

Methane (CH_4) is chemically unstable in the current oxidizing atmosphere of Mars. It would quickly break down due to ultraviolet radiation from the Sun and chemical reactions with other gases. Therefore, a persistent presence of methane in the atmosphere may imply the existence of a source to continually replenish the gas.

Trace amounts of methane, at the level of several parts per billion (ppb), were first reported in Mars' atmosphere by a team at the NASA Goddard Space Flight Center in 2003. Large differences in the abundances were measured between observations taken in 2003 and 2006, which suggested that the methane was locally concentrated and probably seasonal. On June 7, 2018, NASA announced it has detected a seasonal variation of methane levels on Mars.

The ExoMars Trace Gas Orbiter (TGO), launched in March 2016, began on April 21, 2018 to map the concentration and sources of methane in the atmosphere, as well as its decomposition products such as formaldehyde and methanol. As of May 2019, the Trace Gas Orbiter showed that the concentration of methane is under detectable level (< 0.05 ppbv).

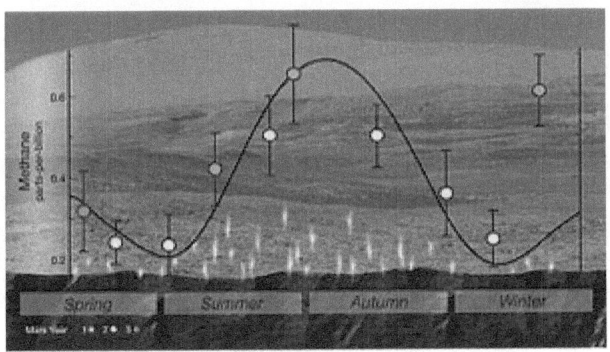

Curiosity detected a cyclical seasonal variation in atmospheric methane.

The principal candidates for the origin of Mars' methane include non-biological processes such as water-rock reactions, radiolysis of water, and pyrite formation, all of which produce H2 that could then generate methane and other hydrocarbons via Fischer–Tropsch synthesis with CO and CO2. It has also been shown that methane could be produced by a process involving water, carbon dioxide, and the mineral olivine, which is known to be common on Mars. Although geologic sources of methane such as serpentinization are possible, the lack of current volcanism, hydrothermal activity or hotspots are not favorable for geologic methane.

Living microorganisms, such as methanogens, are another possible source, but no evidence for the presence of such organisms has been found on Mars, until June 2019 as methane was detected by the Curiosity rover.

Methanogens do not require oxygen or organic nutrients, are non-photosynthetic, use hydrogen as their energy source and carbon dioxide (CO2) as their carbon source, so they could exist in subsurface environments on Mars. If microscopic Martian life is producing the methane, it probably resides far below the surface, where it is still warm enough for liquid water to exist.

Since the 2003 discovery of methane in the atmosphere, some scientists have been designing models and in vitro experiments testing the growth of methanogenic bacteria on simulated Martian soil, where all four methanogen strains tested produced substantial levels of methane, even in the presence of 1.0wt% perchlorate salt.

A team led by Levin suggested that both phenomena—methane production and degradation—could be accounted for by an ecology of methane-producing and methane-consuming microorganisms.

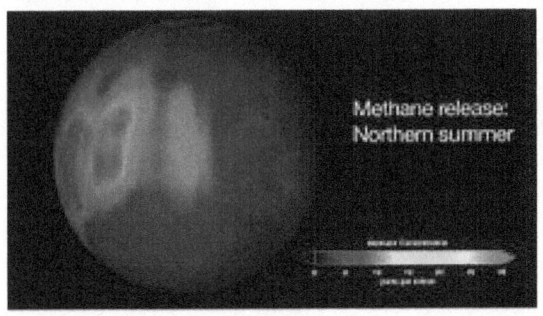

Distribution of methane in the atmosphere of Mars in the Northern Hemisphere during summer

Research at the University of Arkansas presented in June 2015 suggested that some methanogens could survive on Mars' low pressure. Rebecca Mickol found that in her laboratory, four species of methanogens survived low-pressure conditions that were similar to a subsurface liquid aquifer on Mars. The four species that she tested were Methanothermobacter wolfeii, Methanosarci na barkeri, Methanobacterium formicicum, and Methanococcus maripaludis.

In June 2012, scientists reported that measuring the ratio of hydrogen and methane levels on Mars may help determine the likelihood of life on Mars. According to the scientists, "low H_2/CH_4 ratios (less than approximately 40)" would "indicate that life is likely present and active". The observed ratios in the lower Martian atmosphere were "approximately 10 times" higher "suggesting that biological processes may not be responsible for the observed CH_4". The scientists suggested measuring the H_2 and CH_4 flux at the Martian surface for a more accurate assessment. Other scientists have recently reported methods of detecting hydrogen and methane in extraterrestrial atmospheres.

Even if rover missions determine that microscopic Martian life is the seasonal source of the methane, the life forms probably reside far below the surface, outside of the rover's reach.

Formaldehyde

In February 2005, it was announced that the Planetary Fourier Spectrometer (PFS) on the European Space Agency's Mars Express Orbiter had detected traces of formaldehyde in the atmosphere of Mars. Vittorio Formisano, the director of the PFS, has speculated that the formaldehyde could be the byproduct of the oxidation of methane and, according to him, would provide evidence that Mars is either extremely geologically active or harboring colonies of microbial life. NASA scientists consider the preliminary findings well worth a follow-up but have also rejected the claims of life.

Viking lander biological experiments

The 1970s the Viking program placed two identical landers on the surface of Mars tasked to look for bio signatures of microbial life on the surface. Of the four experiments performed by each Viking lander, only the 'Labeled Release' (LR) experiment gave a positive result for metabolism, while the other three did not detect organic compounds. The LR was a specific experiment designed to test only a narrowly defined critical aspect of the theory concerning the possibility of life on Mars; therefore, the overall results were declared inconclusive. No Mars lander mission has found meaningful traces of biomolecules or bio signatures. The claim of extant microbial life on Mars is based on old data collected by the Viking landers, currently reinterpreted as sufficient evidence of life, mainly by Gilbert Levin, Joseph D. Miller, Navarro, Giorgio Bianciardi and Patricia Ann Straat, that the Viking LR experiments detected extant microbial life on Mars.

Assessments published in December 2010 by Rafael Navarro–Gonzáles indicate that organic compounds "could have been present" in the soil analyzed by both Viking 1 and 2. The study determined that perchlorate—discovered in 2008 by Phoenix lander—can destroy organic compounds when heated, and produce chloromethane and dichloromethane as a byproduct, the identical chlorine compounds discovered by both Viking landers when they performed the same tests on Mars. Because perchlorate would have broken down any Martian organics, the question of whether or not Viking found organic compounds is still wide open.

The Labeled Release evidence was not generally accepted initially, and, to this day lacks the consensus of the scientific community.

Curiosity rover sediment sampling

In June 2018, NASA reported that the Curiosity rover had found evidence of complex organic compounds from mudstone rocks aged approximately 3.5 billion years old, sampled from two distinct sites in a dry lake in the Pahrump Hills of the Gale crater. The rock samples, when pyrolyzed via the Curiosity's Sample Analysis at Mars instrument, released an array of organic molecules; these include sulfur-containing thiophenes, aromatic compounds such as benzene and toluene, and aliphatic compounds such as propane and butene.

The concentration of organic compounds is 100-fold higher than earlier measurements. The authors speculate that the presence of sulfur may have helped preserve them. The products resemble those obtained from the breakdown of kerogen, a precursor to oil and natural gas on Earth. NASA stated that these findings are not evidence that life existed on the planet, but that the organic compounds needed to sustain microscopic life were present and there may be deeper sources of organic compounds on the planet.

Meteorites

As of 2018, there are 224 known Martian meteorites (some of which were found in several fragments). These are valuable because they are the only physical samples of Mars available to Earth-bound laboratories. Some researchers have argued that microscopic morphological features found in ALH84001 are biomorphs, however this interpretation has been highly controversial and is not supported by the majority of researchers in the field.

Seven criteria have been established for the recognition of past life within terrestrial geologic samples. Those criteria are:

1. Is the geologic context of the sample compatible with past life?

2. Is the age of the sample and its stratigraphic location compatible with possible life?

3. Does the sample contain evidence of cellular morphology and colonies?

4. Is there any evidence of biominerals showing chemical or mineral disequilibria?

5. Is there any evidence of stable isotope patterns unique to biology?

6. Are there any organic biomarkers present?
7. Are the features indigenous to the sample?

For general acceptance of past life in a geologic sample, essentially most or all of these criteria must be met. All seven criteria have not yet been met for any of the Martian samples.

ALH8400

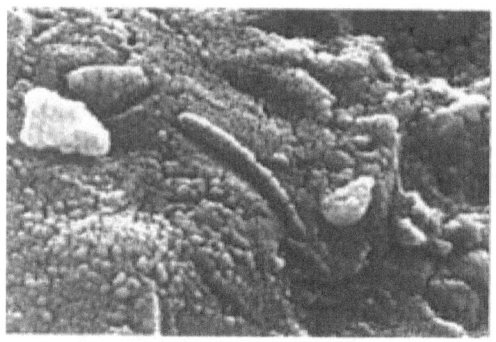

An electron microscope reveals bacteria-like structures in meteorite fragment ALH84001

In 1996, the Martian meteorite ALH84001, a specimen that is much older than the majority of Martian meteorites that have been recovered so far, received considerable attention when a group of NASA scientists led by David S. McKay reported microscopic features and geochemical anomalies that they considered to be best explained by the rock having hosted Martian bacteria in the distant past. Some of these features resembled terrestrial bacteria, aside from their being much smaller than any known form of life. Much controversy arose over this claim, and ultimately all of the evidence McKay's team cited as evidence of life was found to be explainable by non-biological processes. Although the scientific community has largely rejected the claim ALH 84001 contains evidence of ancient Martian life, the controversy associated with it is now seen as a historically significant moment in the development of exobiology.

Nakhla meteorite

Nakhla

The Nakhla meteorite fell on Earth on June 28, 1911, on the locality of Nakhla, Alexandria, Egypt.

In 1998, a team from NASA's Johnson Space Center obtained a small sample for analysis. Researchers found preterrestrial aqueous alteration phases and objects of the size and shape consistent with Earthly fossilized nanobacteria. Analysis with gas chromatography and mass spectrometry (GC-MS) studied its high molecular weight polycyclic aromatic hydrocarbons in 2000, and NASA scientists concluded that as much as 75% of the organic compounds in Nakhla "may not be recent terrestrial contamination".

This caused additional interest in this meteorite, so in 2006, NASA managed to obtain an additional and larger sample from the London Natural History Museum. On this second sample, a large dendritic carbon content was observed. When the results and evidence were published in 2006, some independent researchers claimed that the carbon deposits are of biologic origin. It was remarked that since carbon is the fourth most abundant element in the Universe, finding it in curious patterns is not indicative or suggestive of biological origin.

Shergotty

The Shergotty meteorite, a 4 kg Martian meteorite, fell on Earth on Shergotty, India on August 25, 1865, and was retrieved by witnesses almost immediately. It is composed mostly of pyroxene and thought to have undergone preterrestrial aqueous alteration for several centuries. Certain features in its interior suggest remnants of a biofilm and its associated microbial communities.

Yamato 000593

Yamato 000593 is the second largest meteorite from Mars found on Earth. Studies suggest the Martian meteorite was formed about 1.3 billion years ago from a lava flow on Mars. An impact occurred on Mars about 12 million years ago and ejected the meteorite from the Martian surface into space. The meteorite landed on Earth in Antarctica about 50,000 years ago. The mass of the meteorite is 13.7 kg (30 lb) and it has been found to contain evidence of past water movement. At a microscopic level, spheres are found in the meteorite that are rich in carbon compared to surrounding areas that lack such spheres. The carbon-rich spheres may have been formed by biotic activity according to NASA scientists.

All About Mars Journeys and Settlement

13.0 Strange Things Seen on Mars

There are some very strange objects our rovers and satellites have seen on Mars. Are these real or do we just think so from our over active imaginations?

Faces on Mars

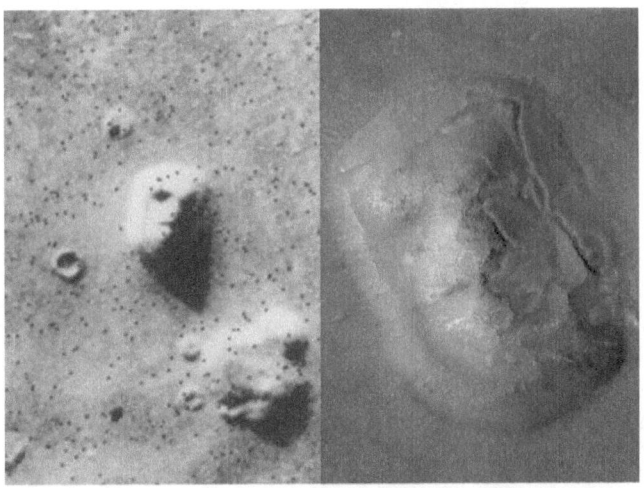

Cydonia is a region on the planet Mars that has attracted both scientific and popular interest. The name originally referred to the albedo feature that was visible from earthbound telescopes.

The area borders the plains of Acidalia Planitia and the highlands of Arabia Terra. The region includes the named features Cydonia Mensae, an area of flat-topped mesa-like features; Cydonia Colles, a region of small hills or knobs; and Cydonia Labyrinthus, a complex of intersecting valleys. As with other albedo features on Mars, the name Cydonia was drawn from classical antiquity, in this case from Kydonia a historic polis (city state) on the island of Crete. Cydonia contains the "Face on Mars", located about halfway between the craters Arandas and Bamberg.

Later images showed this original face image was not correct.

Statue on Mars

Crab Like Animal On Wall

Trees on Mars

A Possible Petrified Animal

A petrified Snake?

14.0 Sci Fi and Fantasy Books About Mars

A number of science fiction and fantasy books have been written about Mars over the years. Here are a few of the ones which made a lasting impact on popular culture about Mars:

14.1 Edgar Rice Burroughs

Barsoom is a fictional representation of the planet Mars created by American pulp fiction author Edgar Rice Burroughs. The first Barsoom tale was serialized as Under the Moons of Mars in 1912, and published as a novel as A Princess of Mars in 1917. Ten sequels followed over the next three decades, further extending his vision of Barsoom and adding other characters. The first five novels are in the public domain in U.S., and the entire series is free around the world on Project Gutenberg Australia, but the books are still under copyright in most of the rest of the world.

The Barsoom series, where John Carter in the late 19th century is mysteriously transported from Earth to a Mars suffering from dwindling resources, has been cited by many well-known science fiction writers as having inspired them. Elements of the books have been adapted by many writers, in novels, short stories, comics, television and film.

14.2 The Mars Series

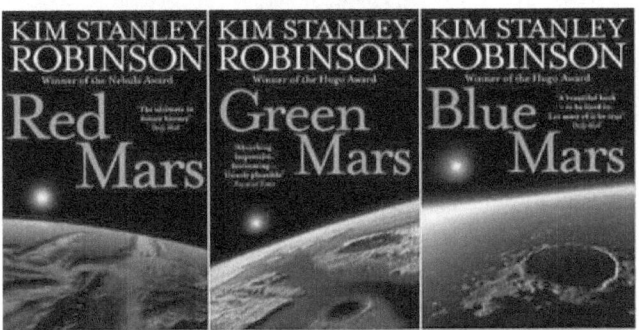

The Mars trilogy is a series of science fiction novels by Kim Stanley Robinson that chronicles the settlement and terraforming of the planet Mars through the personal and detailed viewpoints of a wide variety of characters spanning almost two centuries. Ultimately more utopian than dystopian, the story focuses on egalitarian, sociological, and scientific advances made on Mars, while Earth suffers from overpopulation and ecological disaster.

The three novels are Red Mars (1992), Green Mars (1993), and Blue Mars (1996). The Martians (1999) is a collection of short stories set in the same fictional universe. Red Mars won the BSFA Award in 1992 and Nebula Award for Best Novel in 1993. Green Mars won the Hugo Award for Best Novel and Locus Award for Best Science Fiction Novel in 1994. Blue Mars also won the Hugo and Locus Awards in 1997.

Icehenge (1984), Robinson's first novel about Mars, is not set in this universe but deals with similar themes and plot elements. The trilogy shares some similarities with Robinson's more recent novel 2312 (2012); for instance, the terraforming of Mars and the extreme longevity of the characters in both novels.

14.3 The Martian By Andy Weir

The Martian is a 2011 science fiction novel written by Andy Weir. It was his debut novel under his own name. It was originally self-published in 2011; Crown Publishing purchased the rights and re-released it in 2014. The story follows an American astronaut, Mark Watney, as he becomes stranded alone on Mars in 2035 and must improvise in order to survive. The Martian, a film adaptation directed by Ridley Scott and starring Matt Damon, was released in October 2015.

All About Mars Journeys and Settlement

15.0 Summary

A manned mission to Mars will be the greatest exploration adventure of the twenty first century. It will be comparable to man landing on the Moon in 1969. The effort to get to Mars will also advance many technologies.

The people making this journey will be undergoing a challenge like Christopher Columbus and his crew crossing the Atlantic or Magellan sailing around the world. Man landing on Mars will be a major event in the history of exploration.

In this book I've tried to capture the history of ideas and the challenges which need to be overcome to make this journey possible.

There are even some visionaries who want to build colonies on Mars in the near future. Will they be ahead of their times or courting disaster?

This book is an updated version of my first one about Mars in 2019. There was a lot of new information so I hope you enjoyed it.

I'm looking forward to this adventure.

All the Best,
Martin K. Ettington
July 2022

16.0 Bibliography

1. Human Mission tos Mars. https://en.wikipedia.org/wiki/Human_mission_to_Mars. [Online]

2. Mars Cycler. https://en.wikipedia.org/wiki/Mars_cycler. [Online]

3. Nuclear Thermal Rocket. https://en.wikipedia.org/wiki/Nuclear_thermal_rocket. [Online]

4. The Mars Colony Challenge. https://www.space.com/41697-hp-mars-colony-challenge.html. [Online]

5. Overview of Mars MIssions. https://en.wikipedia.org/wiki/Exploration_of_Mars#Overview_of_missions. [Online]

6. Mars Mission Astronaut Dangers. https://www.space.com/crewed-mars-mission-astronaut-dangers.html. [Online]

7. All About Mars Facts. https://mars.nasa.gov/all-about-mars/facts/. [Online]

8. SpaceX timeline for getting to Mars. https://www.inverse.com/article/51291-spacex-here-s-the-timeline-for-getting-to-mars-and-starting-a-colony. [Online]

9. Mission to Mars Psychological Effects. https://www.apa.org/monitor/2018/06/mission-mars. [Online]

10. Sublake Settlements. https://www.centauri-dreams.org/2020/05/29/sublake-settlements-for-mars/. [Online]

11. The Homestead Project. https://www.space.com/1419-homestead-project-making-mars-settlement-reality.html. [Online]

12. Department, Nuclear Engineering. MISSION TO MARS:HOW TO GET PEOPLE THEREAND BACK WITH NUCLEAR ENERGY. 2003.

13. The Long Road to Mars. https://physicsworld.com/a/the-long-road-to-mars/. [Online]

14. Life on Mars. https://en.wikipedia.org/wiki/Life_on_Mars. [Online]

15. https://www.planetary.org/space-missions/perseverance. Perseverence. [Online] 2021.

16. https://www.energy.gov/ne/articles/6-things-you-should-know-about-nuclear-thermal-propulsion. 6 things to know about nuclear thermal propulsion. [Online] 2022.

17. https://techcrunch.com/2020/04/03/nasa-details-how-it-plans-to-establish-a-sustained-human-presence-on-the-moon/. NASA Details Plans to Establish a Human Presence on Mars. [Online] 2020.

17.0 Index